Structure and Phase Transformations in Thin Films

Structure and Phase Transformations in Thin Films

Editor

Zsolt Czigány

Basel • Beijing • Wuhan • Barcelona • Belgrade • Novi Sad • Cluj • Manchester

Editor
Zsolt Czigány
Centre for Energy Research
Budapest
Hungary

Editorial Office
MDPI
St. Alban-Anlage 66
4052 Basel, Switzerland

This is a reprint of articles from the Special Issue published online in the open access journal *Coatings* (ISSN 2079-6412) (available at: https://www.mdpi.com/journal/coatings/special_issues/struct_transit).

For citation purposes, cite each article independently as indicated on the article page online and as indicated below:

Lastname, A.A.; Lastname, B.B. Article Title. *Journal Name* **Year**, *Volume Number*, Page Range.

ISBN 978-3-0365-8734-9 (Hbk)
ISBN 978-3-0365-8735-6 (PDF)
doi.org/10.3390/books978-3-0365-8735-6

© 2023 by the authors. Articles in this book are Open Access and distributed under the Creative Commons Attribution (CC BY) license. The book as a whole is distributed by MDPI under the terms and conditions of the Creative Commons Attribution-NonCommercial-NoDerivs (CC BY-NC-ND) license.

Contents

Zsolt Czigány
Structure and Phase Transformations in Thin Films
Reprinted from: *Coatings* 2023, 13, 1233, doi:10.3390/coatings13071233 1

Mohamed Arfaoui, György Radnóczi and Viktória Kovács Kis
Transformations in CrFeCoNiCu High Entropy Alloy Thin Films during In-Situ Annealing in TEM
Reprinted from: *Coatings* 2020, 10, 60, doi:10.3390/coatings10010060 5

Jenő Gubicza, Garima Kapoor, Dávid Ugi, László Péter, János L. Lábár and György Radnóczi
Micropillar Compression Study on the Deformation Behavior of Electrodeposited Ni–Mo Films
Reprinted from: *Coatings* 2020, 10, 205, doi:10.3390/coatings10030205 21

Tivadar Lohner, Edit Szilágyi, Zsolt Zolnai, Attila Németh, Zsolt Fogarassy, Levente Illés, et al.
Determination of the Complex Dielectric Function of Ion-Implanted Amorphous Germanium by Spectroscopic Ellipsometry
Reprinted from: *Coatings* 2020, 10, 480, doi:10.3390/coatings10050480 35

Faisal I. Alresheedi and James E. Krzanowski
X-ray Diffraction Investigation of Stainless Steel—Nitrogen Thin Films Deposited Using Reactive Sputter Deposition
Reprinted from: *Coatings* 2020, 10, 984, doi:10.3390/coatings10100984 45

Nikolett Hegedüs, Riku Lovics, Miklós Serényi, Zsolt Zolnai, Péter Petrik, Judit Mihály, et al.
Examination of the Hydrogen Incorporation into Radio Frequency-Sputtered Hydrogenated SiN_x Thin Films
Reprinted from: *Coatings* 2021, 11, 54, doi:10.3390/coatings11010054 61

Dung Nguyen Trong, Van Cao Long and Ştefan Ţălu
The Study of the Influence of Matrix, Size, Rotation Angle, and Magnetic Field on the Isothermal Entropy, and the Néel Phase Transition Temperature of Fe_2O_3 Nanocomposite Thin Films by the Monte-Carlo Simulation Method
Reprinted from: *Coatings* 2021, 11, 1209, doi:10.3390/coatings11101209 75

Igor Bychkov, Sergey Belim, Ivan Maltsev and Vladimir Shavrov
Phase Transition and Magnetoelectric Effect in 2D Ferromagnetic Films on a Ferroelectric Substrate
Reprinted from: *Coatings* 2021, 11, 1325, doi:10.3390/coatings11111325 91

Jun Zhang, Lijing Peng, Xiaoyang Wang, Dongling Liu and Nan Wang
Effects of Zr/(Zr+Ti) Molar Ratio on the Phase Structure and Hardness of $Ti_xZr_{1-x}N$ Films
Reprinted from: *Coatings* 2021, 11, 1342, doi:10.3390/coatings11111342 101

Petr Shvets, Ksenia Maksimova and Alexander Goikhman
Raman Spectroscopy of V_4O_7 Films
Reprinted from: *Coatings* 2022, 12, 291, doi:10.3390/coatings12030291 111

Lijing Peng, Jun Zhang and Xiaoyang Wang
Phase Composition, Hardness, and Thermal Shock Properties of AlCrTiN Hard Films with High Aluminum Content
Reprinted from: *Coatings* 2023, 13, 547, doi:10.3390/coatings13030547 119

Editorial

Structure and Phase Transformations in Thin Films

Zsolt Czigány

Centre for Energy Research, Institute of Technical Physics and Materials Science, Konkoly Thege M. út 29-33, H-1121 Budapest, Hungary; czigany.zsolt@ek-cer.hu

1. Introduction

The field of thin films has gone through a great development in recent decades. Besides the multiple applications of thin films, an increasing competence can be observed in tailoring the film microstructure via composition and deposition parameters. The first approach led to the development of, e.g., multicomponent carbide, nitride and oxynitride [1,2] hard coatings and superalloys [3,4], and turned recent interest towards high-entropy alloys [5–8]. The latter approach resulted in significant improvement of deposition techniques where energetic surface bombardment can influence surface mobility and the formed bonding at the growth surface [9,10]. Revealing the microstructure is inevitable for understanding how defects, the ordering of atoms within the crystalline unit cell or the formation of multiple phases and their arrangement influences the performance of coatings on the macroscopic scale. Structural features on the nanoscale may also provide excellent properties to novel nanocomposites and special nanolaminated structures, e.g., MAX phases [11–13]. The achievements that can be quantified in terms of hardness, elasticity, corrosion resistance or adhesion can be utilized in the improved lifetime of coatings and coated parts. However, we know little about the effect of additives and their role in the secondary phase formation and segregation and their influence on macroscopic properties (e.g., corrosion). Moreover, the research is often limited to the formed phases and structures. The existing knowledge should be completed with the exploration of the mechanism of the ongoing processes by their in situ monitoring. The recent development of state-of-the-art in situ TEM specimen holders may boost the available information about the structural transformations due to an elevated temperature, deformation or chemical ambient.

2. Review of Special Issue Contents

This Special Issue [14–23] aimed at attracting cutting-edge research and review articles on the preparation and characterization of materials with excellent properties or the related modelling approaches. The published papers in this Special Issue cover a wide range of topics, including high-entropy alloys [14], hard materials [15,21,23], effects of additives [15,17,18] on structure and properties, and the identification of Magnéli phases [22]. Computer modelling of magnetic properties [19,20] and optical modelling of dielectric function [16] are also represented in this volume.

The first published paper [14] presents an in situ TEM-annealing study of the microstructural evolution of CrFeCoNiCu high-entropy alloy (HEA) thin films. A post-annealing investigation of the samples was carried out using high-resolution transmission electron microscopy and EDS measurements. The film is a structurally stable single-phase FCC HEA up to 400 °C. At 450 °C, the formation of a BCC phase was observed; however, the morphology of the film was preserved. This type of transformation is attributed to diffusionless processes (martensitic or massive), while the fast morphological and structural changes above 550 °C are controlled by volume diffusion processes. The structure of the intermetallic phase formed at high temperatures is modelled as a supercell of the BCC phase.

The second paper [15] shows an example of how additives can influence the mechanical properties of a material. They studied the influence of 0.4–5.3 at% Mo addition on the compression behaviour of electrodeposited Ni films using micropillar deformation tests. They observed a lower grain size and higher dislocation and twin density at high Mo content, the differences of which resulted in a four-times higher yield strength compared to low Mo samples. The strain softening of low Mo samples after repeated deformation was attributed to detwinning during deformation.

In the third paper [16], amorphous Ge (a-Ge) was created in single-crystalline Ge via ion implantation. It was shown that high optical density is available when implanting low-mass Al ions using a dual-energy approach. The optical properties were measured using multiple angle of incidence spectroscopic ellipsometry. The Cody–Lorentz dispersion model was the most suitable and was capable of describing the dielectric function using a few parameters in the wavelength range from 210 to 1690 nm. The results of the optical model were consistent with Rutherford backscattering spectrometry and cross-sectional electron microscopy measurements, including the agreement of the layer thickness within experimental uncertainty. Accurate reference dielectric functions play an important role in the research development and reproducible preparation of optical materials, such as in integrated optics, optoelectronics or photovoltaics.

In the fourth paper [17], the effect of (~28 and 32 at%) N incorporation was investigated using X-ray diffraction on nitrogen-containing 304 stainless steel thin films deposited by reactive RF magnetron sputtering as a function of substrate temperature and bias. The extent of the diffraction anomaly, i.e., relative peak shifts of (111) and (200) peak positions, was determined using a calculated parameter, denoted $R_B = \sin^2\theta_{111}/\sin^2\theta_{200}$. The normal value for R_B for FCC-based structures is 0.75, and R_B increases as the (002) peak shifts closer to the (111) peak. In this study, the R_B values for the deposited films increased with substrate bias but decreased with substrate temperature (but still always >0.75). Since the split of 200 reflection (due to assumed tetragonal distortion) was not observed, a defect-based hypothesis is more viable as an explanation for the diffraction anomaly.

In the fifth paper [18], amorphous hydrogen-free silicon nitride (a-SiN$_x$) and amorphous hydrogenated silicon nitride (a-SiN$_x$:H) films were deposited via radio frequency (RF) sputtering applying various hydrogen flows. The refractive index of 1.96 was characteristic for hydrogen-free SiN$_x$ thin film, and with increasing H$_2$ flow, it decreased to 1.89. Hydrogenation during the sputtering process increased the porosity of the thin films compared with hydrogen-free SiN$_x$. Higher porosity is consistent with a lower refractive index. The hydrogen content of hydrogenated films was 4 at% and 6 at% according to Fourier-transform infrared spectroscopy (FTIR) and elastic recoil detection analysis (ERDA), respectively. The molecular form of hydrogen was released at a temperature of ~65 °C from the film after annealing, while blisters of 100 nm in diameter were formed on the thin film surface. The low activation energy deduced from the Arrhenius method indicated the diffusion of hydrogen molecules.

The sixth paper [19] provides a study of the Néel phase transition temperature of Fe$_2$O$_3$ nanocomposite thin films via Monte-Carlo simulation taking into account a variety of parameters. They found that particle size and magnetic field are the main parameters that determine the temperature of Neél transition (T_{Ntot}). The results also show that in Fe$_2$O$_3$ thin films, T_{Ntot} is always smaller than in the case of Fe$_2$O$_3$ nanoparticles and bulk Fe$_2$O$_3$. For a nanoparticle size of 12 nm, T_{Ntot} = 300 K. Furthermore, there is a linear relationship between T_{Ntot} and nanoparticle size. The results can be utilized in the design of magnetic devices and in biomedical applications.

The seventh paper [20] is another computer simulation modelling of the ferroelectric substrate influence on the condition and magnetic properties of 2D ferromagnetic nanofilms using the two-dimensional Frenkel–Kontorova (FK) potential to simulate the substrate effect on the film. The Ising model and Wolf cluster algorithm are used to describe the magnetic behaviour of an FM film. The results show that uniform deformations of the substrate lead to inhomogeneous deformations of the film. The interaction between the substrate and film causes film deformations leading to superstructures with low and high atomic concentrations. Heating or external electric field can cause substrate deformations, and consequent substrate-induced structural phase transition takes place. The Curie temperature decreases with both substrate compression and stretching.

In the eighth paper [21], the properties of $Ti_xZr_{1-x}N$ hard coatings were investigated. $Ti_xZr_{1-x}N$ films with $Zr/(Zr + Ti)$ molar ratios from 20% to 80% were prepared with multi-arc ion plating using different combinations of either elemental Ti and Zr or TiZr alloy targets. The as-deposited $Ti_xZr_{1-x}N$ films formed substitutional fcc solid solutions with a lattice constant consistent with Vegard's law. When the Zr/Ti molar ratio was 40:60 or 60:40, the films showed (111) and (220) preferred growth orientation, while at other compositions, the films exhibited (111) preferred orientation. These two compositions coincide with the hardness maxima with values over 30 GPa.

In the ninth paper [22], the structure and phase transition of V_4O_7 Magnéli phase was investigated. A thin film of vanadium oxide Magnéli phase V_4O_7 was produced using cathodic arc sputtering of V target at 600 °C and 0.045 Pa of oxygen atmosphere. The stoichiometric composition of V_4O_7 was confirmed using Rutherford backscatter spectrometry. The similarity of structures of Magnéli phases may make it challenging to identify the phase precisely using X-ray diffraction, especially if the sample is nanostructured, highly textured, and strained. The authors proved that Raman spectroscopy can be a sensitive indicator of minor structural changes even at room temperature with the careful tuning of the laser power on the sample. Metal–insulator phase transition in V_4O_7 thin films was observed via Raman spectroscopy between −50 and −55 °C.

In the tenth paper [23], the composition of AlCrTiN quaternary nitride films is optimized. Ternary nitrides, like TiCrN, generally form face-centred cubic solid solutions and have superior properties in terms of phase composition, hardness, and thermal shock resistance and adhesion to the substrate due to the addition of alloying elements. The effect of high Al content was investigated in this work. AlCrTiN hard coatings were prepared via reactive multi-arc ion plating on high-speed steel substrates via the co-deposition of AlCr and AlTi dual-arc source alloy targets in N. The optimal composition of the AlCrTiN hard films is 25:13:15:47 (at%), based on the consideration of hardness, adhesion, and thermal shock cycling resistance. This optimal AlCrTiN hard film can be suggested as an option for protective coatings in applications up to 600 °C.

Funding: National Research Development and Innovation Fund Office, Hungary, Grant Number: K-125100.

Conflicts of Interest: The author declares no conflict of interest.

References

1. Gogotsi, Y.G.; Andrievski, R.A. (Eds.) *Materials Science of Carbides, Nitrides and Borides in NATO Science Series, 3 High Technology–Volume 68*; Springer-Science+Business Media: Dordrecht, The Nederlands, 1999.
2. Rasaki, S.A.; Zhang, B.; Anbalgam, K.; Thomas, T.; Yang, M. Synthesis and application of nano-structured metal nitrides and carbides: A review. *Prog. Solid State Chem.* **2018**, *50*, 1–15. [CrossRef]
3. Moverare, J. (Ed.) Special Issue "Superalloys". Available online: https://www.mdpi.com/journal/metals/special_issues/superalloys (accessed on 4 July 2023).
4. Shahwaz, M.; Nath, P.; Sen, I. A critical review on the microstructure and mechanical properties correlation of additively manufactured nickel-based superalloys. *J. Alloys Compd.* **2022**, *907*, 164530. [CrossRef]
5. Cantor, B.; Chang, I.T.H.; Knight, P.; Vincent, A.J.B. Microstructural development in equiatomic multicomponent alloys. *Mater. Sci. Eng. A* **2004**, *375*, 213–218. [CrossRef]
6. Miracle, D.B.; Senkov, O.N. A critical review of high entropy alloys and related concepts. *Acta Mater.* **2017**, *122*, 448–511. [CrossRef]
7. George, E.P.; Curtin, W.A.; Tasan, C.C. High entropy alloys: A focused review of mechanical properties and deformation mechanisms. *Acta Mater.* **2020**, *188*, 435–474. [CrossRef]
8. Kumari, P.; Gupta, A.K.; Mishra, R.K.; Ahmad, M.S.; Shahi, R.R. A Comprehensive Review: Recent Progress on Magnetic High Entropy Alloys and Oxides. *J. Magn. Magn. Mater.* **2022**, *554*, 169142. [CrossRef]
9. Mattox, D.M. Particle bombardment effects on thin-film deposition: A review. *J. Vac. Sci. Technol. A* **1989**, *7*, 1105–1114. [CrossRef]
10. Aghda, S.K.; Holzapfel, D.M.; Music, D.; Unutulmazsoy, Y.; Mráz, S.; Bogdanovski, D.; Fidanboy, G.; Hans, M.; Primetzhofer, D.; Méndez, A.S.J.; et al. Ion kinetic energy- and ion flux-dependent mechanical properties and thermal stability of (Ti,Al)N thin films. *Acta Mater.* **2023**, *250*, 118864. [CrossRef]

11. Barsoum, M.W.; El-Raghy, T. The MAX Phases: Unique New Carbide and Nitride Materials: Ternary ceramics turn out to be surprisingly soft and machinable, yet also heat-tolerant, strong and lightweight. *Am. Sci.* **2001**, *89*, 334–343. [CrossRef]
12. Barsoum, M.W.; Radovic, M. Elastic and Mechanical Properties of the MAX Phases. *Annu. Rev. Mater. Res.* **2011**, *41*, 195–227. [CrossRef]
13. Lei, X.; Lin, N. Structure and synthesis of MAX phase materials: A brief review. *Crit. Rev. Solid State Mater. Sci.* **2022**, *47*, 736–771. [CrossRef]
14. Arfaoui, M.; Radnóczi, G.; Kis, V.K. Transformations in CrFeCoNiCu High Entropy Alloy Thin Films during In-Situ Annealing in TEM. *Coatings* **2020**, *10*, 60. [CrossRef]
15. Gubicza, J.; Kapoor, G.; Ugi, D.; Péter, L.; Lábár, J.L.; Radnóczi, G. Micropillar Compression Study on the Deformation Behavior of Electrodeposited Ni–Mo Films. *Coatings* **2020**, *10*, 205. [CrossRef]
16. Lohner, T.; Szilágyi, E.; Zolnai, Z.; Németh, A.; Fogarassy, Z.; Illés, L.; Kótai, E.; Petrik, P.; Fried, M. Determination of the Complex Dielectric Function of Ion-Implanted Amorphous Germanium by Spectroscopic Ellipsometry. *Coatings* **2020**, *10*, 480. [CrossRef]
17. Alresheedi, F.I.; Krzanowski, J.E. X-ray Diffraction Investigation of Stainless Steel-Nitrogen Thin Films Deposited Using Reactive Sputter Deposition. *Coatings* **2020**, *10*, 984. [CrossRef]
18. Hegedüs, N.; Lovics, R.; Serényi, M.; Zolnai, Z.; Petrik, P.; Mihály, J.; Fogarassy, Z.; Balázsi, C.; Balázsi, K. Examination of the Hydrogen Incorporation into Radio Frequency-Sputtered Hydrogenated SiNx Thin Films. *Coatings* **2021**, *11*, 54. [CrossRef]
19. Trong, D.N.; Long, V.C.; Talu, S. The Study of the Influence of Matrix, Size, Rotation Angle, and Magnetic Field on the Isothermal Entropy, and the Néel Phase Transition Temperature of Fe_2O_3 Nanocomposite Thin Films by the Monte-Carlo Simulation Method. *Coatings* **2021**, *11*, 1209. [CrossRef]
20. Bychkov, I.; Belim, S.; Maltsev, I.; Shavrov, V. Phase Transition and Magnetoelectric Effect in 2D Ferromagnetic Films on a Ferroelectric Substrate. *Coatings* **2021**, *11*, 1325. [CrossRef]
21. Zhang, J.; Peng, L.; Wang, X.; Liu, D.; Wang, N. Effects of Zr/(Zr+Ti) Molar Ratio on the Phase Structure and Hardness of $Ti_xZr_{1-x}N$ Films. *Coatings* **2021**, *11*, 1342. [CrossRef]
22. Shvets, P.; Maksimova, K.; Goikhman, A. Raman Spectroscopy of V_4O_7 Films. *Coatings* **2022**, *12*, 291. [CrossRef]
23. Peng, L.; Zhang, J.; Wang, X. Phase Composition, Hardness, and Thermal Shock Properties of AlCrTiN Hard Films with High Aluminum Content. *Coatings* **2023**, *13*, 547. [CrossRef]

Disclaimer/Publisher's Note: The statements, opinions and data contained in all publications are solely those of the individual author(s) and contributor(s) and not of MDPI and/or the editor(s). MDPI and/or the editor(s) disclaim responsibility for any injury to people or property resulting from any ideas, methods, instructions or products referred to in the content.

Article

Transformations in CrFeCoNiCu High Entropy Alloy Thin Films during In-Situ Annealing in TEM

Mohamed Arfaoui [1,2,*], György Radnóczi [1] and Viktória Kovács Kis [1,3]

1. Centre for Energy Research of the Hungarian Academy of Sciences, Konkoly-Thege M. u. 29-33, H-1121 Budapest, Hungary; radnoczi.gyorgy@energia.mta.hu (G.R.); kis.viktoria@energia.mta.hu (V.K.K.)
2. Deptartment of Physics, Eötvös Loránd University, Pázmány Péter sétány 1a, H-1117 Budapest, Hungary
3. Institute of Environmental Sciences, University of Pannonia, Egyetem str. 10, H-8200 Veszprém, Hungary
* Correspondence: Mohamed.arfaoui@mfa.kfki.hu; Tel.: +36-70-244-3026

Received: 13 December 2019; Accepted: 7 January 2020; Published: 9 January 2020

Abstract: In-situ TEM-heating study of the microstructural evolution of CrFeCoNiCu high entropy alloy (HEA) thin films was carried out and morphological and phase changes were recorded. Post annealing investigation of the samples was carried out by high resolution electron microscopy and EDS measurements. The film is structurally and morphologically stable single phase FCC HEA up to 400 °C. At 450 °C the formation of a BCC phase was observed, however, the morphology of the film remained unchanged. This type of transformation is attributed to diffusionless processes (martensitic or massive). From 550 °C fast morphological and structural changes occur, controlled by volume diffusion processes. Fast growing of a new intermetallic phase is observed which contains mainly Cr and has large unit cell due to chemical ordering of components in <100> direction. The surface of the films gets covered with a CrO-type layer, possibly contributing to corrosion resistance of these.

Keywords: high entropy alloy; in-situ TEM annealing; thermal stability; diffusionless transformation; planar disorder; oxide formation

1. Introduction

High entropy alloys (HEAs) are multi-component alloys in which at least five elements of concentration between 5 and 30 at % are combined [1]. The basic idea is to reduce the Gibbs free energy through maximizing the configuration entropy, which facilitates the formation of random solid solution rather than a complex microstructure built up by intermetallic compounds [2,3]. Thus, studying the structure and stability of HEA during high temperature treatments [4–6] both at the micro- and atomic levels, is a critical issue in understanding of their growth processes.

Detailed structural investigations of bulk HEA have shown that some ordered solid solutions as well as intermetallic compounds can precipitate from the homogeneous matrix at room temperature (RT) and at elevated temperatures as well. For example, ordered FCC phases were found at RT in the $Al_{0.5}CoCrCuFeNi$ [7] and $Al_{0.3}CoCrFeNi$ [8] as well disordered BCC and ordered B2 phases in $Al_{0.5}CoCrCuFeNi$ [9] and $Al_xCoCrFeNi$ [10]. The size and the lattice parameter of these BCC grains have been found to be of the order of ten micrometers and 0.289 nm respectively, regardless of their composition.

Several studies deal with structural changes in HEAs at elevated temperatures. The CrFeMnCoNi alloy, considered as a stable HEA at temperatures below its melting point, formed both carbide and σ phase at grain boundaries of the FCC matrix after prolonged exposures to 700 °C [11]. Recent investigations show that in the same material three different phases (L10-NiMn, B2-FeCo and a Cr-rich body-centered cubic, BCC phase) can precipitate at the grain boundaries after long time annealing in the 500–900 °C temperature interval [12]. With a minor addition of Al to CrFeCoNi HEA it also became unstable leading to the formation of a Cr-rich phase after a long time annealing at 750 °C [13]. The formation of σ phase is reported after annealing at 1000 °C for 15 min of the CrCuCoFeNi alloy as well [14]. Synchrotron XRD results [15,16] show that the microstructure at RT of bulk as-cast CrFeCoNiCu alloy exhibits two FCC phases of lattice constants 0.361 nm and 0.358 nm. One of these phases is Cu-poor with dendritic morphology and the other is a Cu-rich interdendritic phase. Moreover, the two-phase structure is preserved even after a heat treatment at 1100 °C and 1250 °C and, in parallel, the Cu content of the interdendritic region increased [17,18].

Most of these results were obtained with the aim of understanding structural changes within the as cast or homogenized HEA at elevated temperatures, and indicate that the effect of sluggish cooperative diffusion and negative mixing entropy of elements, which inhibits the growth of new phases and nanoparticles at low temperatures, diminishes under heat treatment.

Based on these experimental results, scientific attention has been focused on the application of HEAs as thin film materials. Due to the higher cooling rate and vapor-solid formation mechanism [15], these films can provide different (nano) structural properties compared to bulk HEA. The $Al_{0.5}$CrFeCoNiCu [19] thin film reported by Chen et al. in 2004 showed a FCC single solid solution based on X-ray diffraction. According to TEM, the CrFeCoNiCu (200 nm thick HEA thin film) grows in face centered cubic solid solution and no obvious Cu segregation was observed even on the nanometer scale [15,20]. This supports the great potential of the sputtering technique to produce uniform elemental distribution leading to single phase nanostructure due to fast quenching rate as compared to the as-cast samples.

Regarding structural changes of HEA thin films at elevated temperatures only a few in-situ experiments have been published. For example, in-situ X-ray diffraction studies were carried out on 1 μm thick AlCoCrCuFeNi thin films in the temperature range 110–810 °C by Dolique et al. [21]. They found that AlCoCrCuFeNi films were stable up to 510 °C, which indicates an inferior thermal stability to their bulk counterpart (800 °C). The effect of Al on thermal behavior of Al_xCoCrFeNi film was investigated by in-situ TEM heat treatment at 500 °C and 900 °C as well [22]. The results clearly confirmed that increasing the Al content initiates the formation of B2/BCC phases, which are precursors of a sigma phase.

The in-situ in TEM heating and electron irradiation experiment of a 100 nm thick sputtered CrFeCoNiCu HEA thin film in the temperature range 20–720 °C [23] also supports the formation of BCC/B2 phases at 300 °C, which is followed by sigma phase formation at 720 °C. However, a recent, in-situ TEM heating of the same polycrystalline alloy confirmed microstructure stabilization up to 300 °C during 1800 s [24]. On the other hand, it is also known, that the atomic structure (FCC or BCC or both phases) in these alloys play a sensitive role in the mechanical properties [16,18] as well in the nucleation pathway of the sigma phase [10,22,23]. At the same time, the early stages of phase evolution and transformation mechanisms leading to the formation of ordered intermetallic phases (e.g., L12, B1, and sigma) at intermediate and elevated temperatures in multicomponent films are still not revealed in details.

The present work is motivated by the need for a better understanding and controlling the initial stages of phase and compositional instability of as deposited CrFeCoNiCu HEA thin films with increasing temperature. To achieve this goal in-situ TEM-heating study of the microstructural evolution of CrFeCoNiCu thin film is carried out at intermediate and elevated temperatures up to 700 °C to record the morphological and phase changes from as early stages as possible.

2. Materials and Methods

CrFeCoNiCu high entropy (HEA) films were deposited in a high vacuum system by direct current magnetron sputtering. Films with a thickness of 50 nm were grown at room temperature on cleaved NaCl substrate coated with about 5 nm SiO_x layer as well as on 30 nm thick amorphous SiO_x film substrate supported by Cu micro-grids. The substrates were placed at a distance of 120 mm from the target in the centre of the rotating substrate holder. The equiatomic concentration arc melted

CrFeCoNiCu target of 99.95%, purity was mounted 25° toward the vertical. The background pressure of the deposition system was 9×10^{-6} Pa. High purity Ar was used as sputtering gas at a pressure of 0.25 Pa. The target was pre-sputtered for 5 min before deposition with the shutter closed. The DC power and the deposition time were set to 50 W and 5 min respectively resulting in a 50 nm thick film. The films grown on the NaCl/SiO$_x$ substrate were floated off from NaCl and placed on a micro grid. The in-situ annealing was carried out in a Philips CM20 transmission electron microscope, operated at 200 keV. The temperature was raised in steps of 50 °C from room temperature up to 700 °C, having the temperature constant at each step for 5 min. The vacuum in the microscope specimen area was maintained at 4.5×10^{-5} Pa during the whole in-situ annealing process.

Two kinds of in-situ annealing experiments were carried out. First, the changes were followed by selected area diffraction (SAED), recording diffraction patterns from the same area of 1 μm in diameter at the end of each temperature step. For the evaluation of the SAED patterns, the camera constant was calibrated using a 50 nm thick self-supporting random nanocrystalline Al thin film.

In the second case bright field (BF) images were recorded at magnification of 50,000× at the end of each temperature step. The experiments were performed in the way to ensure the observation of changes preferably in the same area of the film. In both in-situ runs (SAED and BF), the sample was quenched to room temperature after having it hold at 700 °C for 5 min.

Energy dispersive X-ray spectroscopy (EDS) (Themis 200 G3 Super-X EDS detector, Eindhoven, The Netherlands) analysis was performed to verify the composition of the samples before and after annealing. Further structural and EDS characterization of the 700 °C annealed sample was carried out by a FEI-Themis transmission electron microscope (Themis 200 G3, Eindhoven, The Netherlands) with a Cs corrected objective lens, in both HREM (High Resolution Electron Microscope) and STEM (Scanning Transmission Electron Microscope) mode (point resolution is around 0.09 nm in HREM mode and 0.16 nm in STEM mode) operated at 200 kV.

3. Results

3.1. Morphology and Diffraction Analysis of the Film during Heating

As can be seen in Figure 1, the electron diffraction pattern of the as-deposited film indicates the presence of a single FCC solid solution phase while the image shows a fine-grained microstructure of about 5–10 nm average grain size. The crystallographic orientation of the grains is random as verified by tilting experiments: the intensity distribution along the diffraction rings does not change up to 35° of tilt angle. By averaging the lattice constants from all measurable FCC rings, the size of the unit cell of the FCC-phase is calculated to be 0.360 nm, very close to that of pure FCC Cu, and is in agreement with values published for the same alloy in literature [15,20,23].

The initial processes taking place during in-situ annealing are the target of the present investigations. Selected stages of the transformation of the CrFeCoNiCu film as followed by SAED and TEM bright field (BF) images are shown in Figure 2.

From the analysis of the SAED patterns (Figures 1a and 2a) we conclude that up to 400 °C only the as grown FCC phase could be observed, and accordingly, BF images (Figure 2b) indicate that the same polycrystalline structure with a grain size about 5–10 nm was preserved. Thus, it could be concluded that the film was structurally stable up to 400 °C.

At 450 °C a new diffraction ring appears in the electron diffraction pattern. Careful analysis of the SAED pattern allowed identifying further diffraction rings which correspond to a BCC phase of lattice parameter $a_{BCC} = 0.294 \pm 0.010$ nm. So, we could conclude that at 450 °C the CrFeCoNiCu HEA film had a two-phase structure, composed from the original FCC and the newly formed BCC phase. It is important to note that these two lattices were related to each other by fulfilling the following relation: $d(111)_{FCC} = d(110)_{BCC}$, i.e., the lattice spacing of the {111} planes in the FCC phase was equal to the lattice spacing of the {110} planes of the BCC phase. No morphological changes were detectable; also the BCC grains could not be identified based on their morphology.

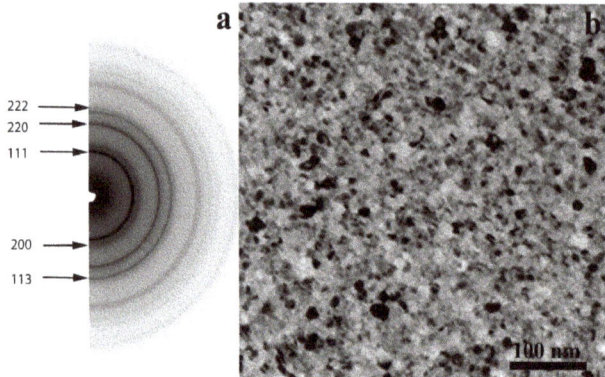

Figure 1. Electron diffraction pattern (**a**) and bright field TEM image (**b**) of the as-deposited CrFeCoNiCu high entropy alloy (HEA) film. The electron diffraction pattern indicates a single phase FCC structure.

Upon further heating, up to 550 °C a few grains with somewhat larger size appear randomly distributed over all of the film area. The diffraction rings of the FCC host phase show texture-like redistribution manifested by the appearance of higher intensity arcs on the rings. At this temperature only the original FCC HEA phase changes, no similar texture-like redistribution was observed in the diffraction rings of the BCC phase.

At 600 °C a drastic change in the microstructure occurs. New grains with larger size appear as marked by A in Figure 2b and grow rapidly to a size of several 100 nm at the expense of the existing two-phase matrix. These grains must belong to the already existing BCC phase as the diffraction rings belonging to the BCC phase (Figure 2a, 600 °C) became discontinuous, indicating grain size growth and, consequently, fewer and larger BCC grains in the same selected area. Nevertheless, at 600 °C the diffraction pattern contains only the FCC and BCC reflections, i.e., no new phases formed. On the other hand, also at this temperature a redistribution of the film material could be observed. Thin, apparently void-like areas formed and grew in the FCC–BCC matrix (B in Figure 2b).

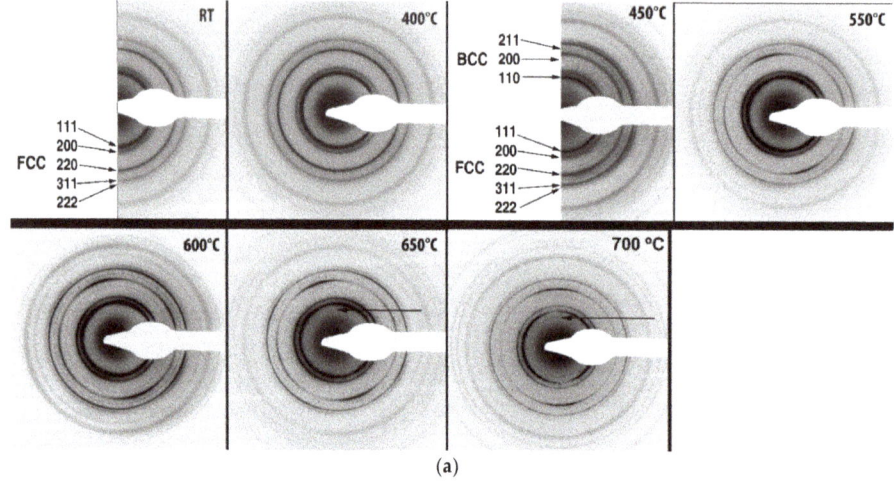

(**a**)

Figure 2. *Cont.*

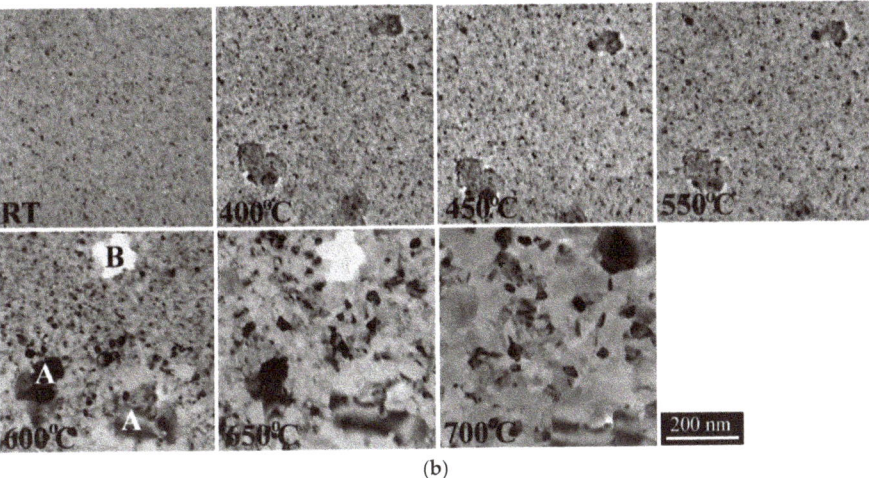

Figure 2. Diffraction patterns (**a**) and in-situ TEM bright field images (**b**) recorded at temperatures marked in individual images.

The number of the large grains increased further at 650 °C (Figures 2b and 3). In agreement with this observation the BCC rings in the diffraction pattern contained more and stronger spots. Some growth of the matrix (FCC) grains could also be concluded from Figure 2b, however, the FCC rings in the diffraction pattern (Figure 2a, 650 °C) remained continuous, indicating only minor changes. In addition, many new reflections appeared including a diffraction ring at 0.25 nm (marked by arrows in Figure 2a).

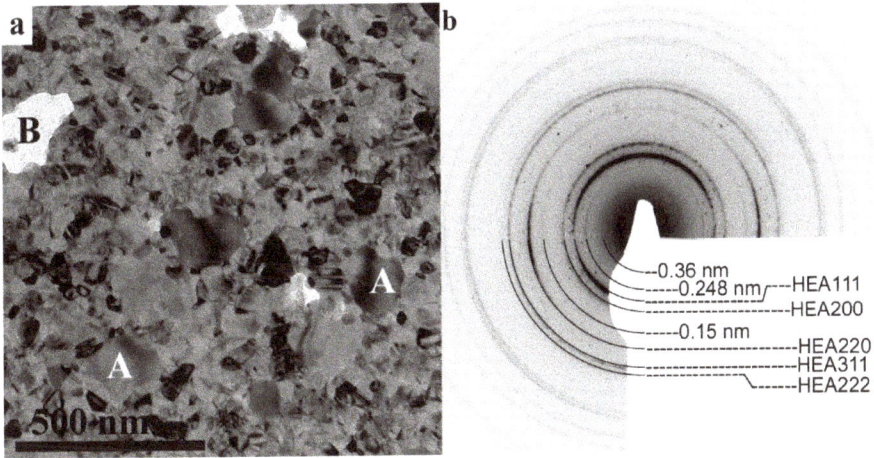

Figure 3. (**a**) Bright field (BF0 TEM image and (**b**) selected area diffraction (SAED) pattern of a 1μm in diameter area (**b**) of the HEA film, recorded at 700 °C.

At 700 °C the scattered reflections already observed at 650 °C increase in number and strength and the diffraction rings of the BCC phase (a = 0.294 nm) became even less continuous though still can be distinguished. This observation implies further grain growth of the A type grains (BCC phase) in the 650–700 °C temperature interval (Figures 2 and 3). On the other hand, the diffraction rings of the FCC phase became also spotted (Figure 3), indicating a grain growth or/and decrease in number of the FCC grains. As a result, the film had a bimodal grain size distribution (Figure 3a). The grains of the matrix phases (mainly FCC) and the large grains (A in Figure 3a) were present having a grain size of 20–50 nm and of about a few hundred nm, respectively.

According to the phase analysis of the diffraction pattern recorded at 700 °C (Figure 3b) the L12 simple cubic phase reflections of the same lattice parameter as the host HEA FCC phase (a = 0.360 nm) were identified besides of the BCC phase (a = 0.294 nm). Moreover, another small grain size FCC phase of a = 0.420 nm was present, which could be identified as a complex oxide phase [25].

The measured and calculated lattice spacing values of the identified phases are summarized in Table 1. As seen from Table 1, there were a lot of overlaps between lattice distances of the four identified phases.

Table 1. Lattice distances of phases measured in the HEA film after annealing at 700 °C for 5 min (Figure 3) and calculated values for the expected phases.

Measured d-Values (nm)	Possible Phases							
	L12 (a = 0.360)		BCC (a = 0.294)		FCC-Oxide (a = 0.420)		HEA FCC (a = 0.360)	
	d-Value	hkl	d-Value	hkl	d-Value	hkl	d-Value	hkl
0.360	0.360	100						
0.248	0.255	110			0.242	111		
0.209	0.208	111	0.208	110	0.210	200	0.208	111
0.180	0.180	200					0.180	220
	0.161	210						
0.150	0.147	211	0.147	200	0.148	220		
0.126	0.127	220			0.127	311	0.127	220
	0.12	221	0.12	211	0.121	222		
0.110	0.109	311			0.105	400	0.109	311
0.100	0.104	222	0.104	220	0.096	331	0.104	222

The textured reflections we considered to belong to the FCC and L12 phases. In this diffraction pattern the BCC phase could not be clearly observed, its scattered reflections became hardly visible in the 650–700 °C temperature interval, the overlaps with the FCC and L12 phases were evident (Table 1). All the rings especially at larger d values were composed of reflections belonging to slightly different spacings. This could be due to composition and consequently lattice parameter fluctuations in the alloy. Nevertheless, the diffraction patterns in Figure 2a unambiguously show, that the 0.150 nm ring cannot belonged to the BCC and L12 phases only, as at 700 °C it reappeared and it was composed from much smaller grains than the L12 and BCC phases had (continuous ring versus spotty rings). So, together with the 0.248 nm ring, which had the same character, the 0.150 nm ring must belong to the FCC oxide phase, the small grain nature of which would be seen in the HREM images below.

3.2. Microstructure Analysis of the Film Annealed at 700 °C

A detailed high resolution structure and composition analysis of the film, formed during in-situ annealing to 700 °C was carried out in a FEI Titan-Themis microscope at room temperature.

The High Angle Angular Dark Field (HAADF) image and elemental maps of the annealed film are shown in Figures 4 and 5. Figure 4 shows grains with dark gray contrast having the lowest average Z in their composition. These grains correspond to dark grains marked A in Figure 2b (600 °C) and 3 as well as in Figure 4. These grains, formed in the temperature range between 600 and 700 °C were embedded in a lighter matrix of higher Z.

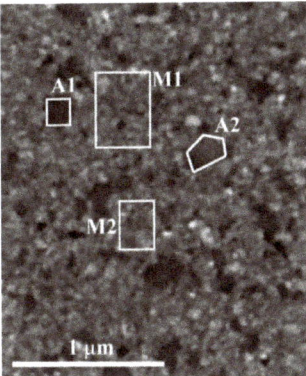

Figure 4. High Angle Angular Dark Field (HAADF) image of the HEA film annealed at 700 °C, measured at room temperature. The frames show the areas used for elemental composition analysis.

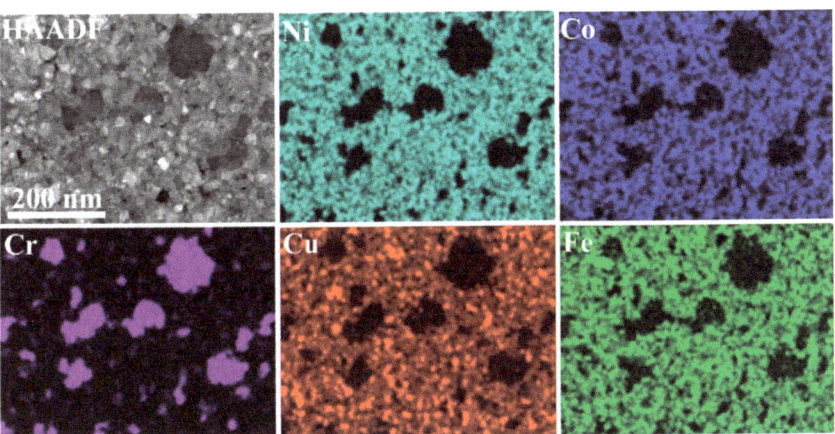

Figure 5. Representative HAADF-STEM image of CrFeCoNiCu alloy film annealed at 700 °C, and its EDS elemental maps showing the distribution of Cr, Fe, Co, Cu, and Ni in the film.

The chemical composition of different morphologies can be found in Table 2. The small grained areas are close in composition to the original HEA with some deficit of Cr and surplus of Cu in them (locations M1 and M2 in Table 2 and Figure 4). Locations A1 and A2 in Table 2 and Figure 4 correspond to dark gray grains and contain mainly Cr (about 63 at. %), while Ni and Co content was significantly less, as indicated by elemental maps in Figure 5. These results show that separation of the components in the originally single phase, homogeneous solid solution HEA film took place during the annealing above 600 °C.

Besides that, the Cr map (Figure 5) shows that smaller grains, with diameter 20–50 nm could also be enriched in Cr. Fe, Co, and Ni were distributed rather evenly in the remaining matrix (Figure 5). The Cu distribution, however, proved to be inhomogeneous on 10–20 nm scale. The elemental maps also indicate that the Cr-rich grains were depleted in Fe, Co, Ni, and Cu, as the Cr map and maps of the other four metal components were complementary.

A general HRTEM view of the structure is shown in Figure 6a. In this figure the remaining mainly FCC HEA metallic phase was seen having a grain size of 20–50 nm, and larger grains (marked A in Figures 3a and 6a) of the Cr-rich phase were present.

Table 2. Chemical composition of the HEA film annealed at 700 °C in areas marked in Figure 4 and measured by energy dispersive X-ray spectroscopy (EDS) in STEM.

Z	Element	Family	Area in Figure 4			
			A1	A2	M1	M2
			Atomic Fraction (%)			
24	Cr	K	62.39	63.29	15.06	10.27
26	Fe	K	12.93	13.86	17.63	19.11
27	Co	K	6.46	9.09	18.57	18.18
28	Ni	K	4.10	1.28	19.28	24.22
29	Cu	K	14.12	12.47	29.47	28.22

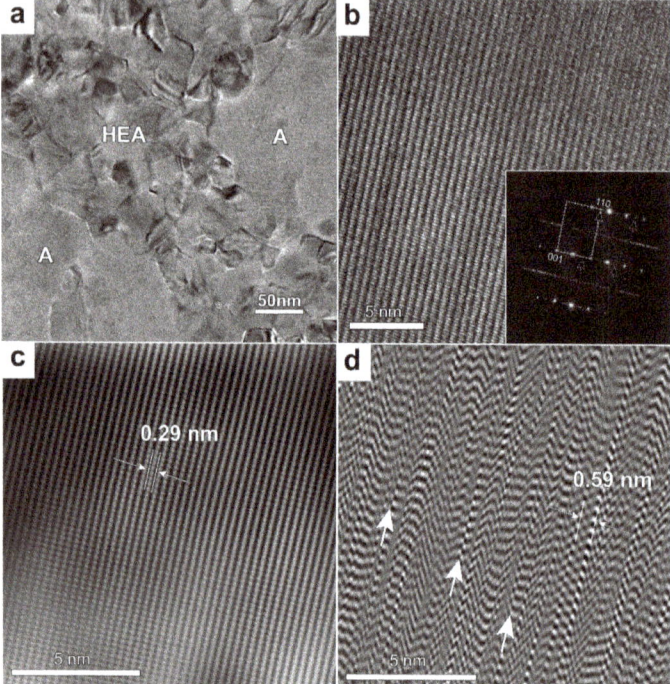

Figure 6. (**a**) Low magnification image showing spatial relationship of the Cr-rich intermetallic phase (A) and the residual HEA phase formed during in-situ annealing up to 700 °C. (**b**) Lattice resolution image of the Cr-rich phase. On the Fourier transform (lower right corner) two directions, corresponding to the unit cell of BCC phase (with $a_0 = 0.294$ nm) are indicated. Note the diffuse scattering parallel to [001]*. Black arrows indicate 1.18 nm periodicity, which is four times the lattice parameter (0.294 nm) of the BCC lattice. (**c**) Enlarged Fourier filtered image made with the discrete (001 and 110 in BCC notation) Fourier components of (**b**), (**d**) Enlarged Fourier filtered image made with the $\frac{1}{2}\frac{1}{2}l$ diffuse scattering (continuous line) on (**b**). White arrows indicate non-periodic fringes parallel to (001). Note the presence of the 0.588 nm periodicity in (**d**) as well.

The internal structure in the large grains displays a striped contrast, corresponding to a large lattice spacing in them. The HRTEM image of these crystallites (grains) and the fast Fourier transform (FFT) diffraction pattern in the insert is shown in Figure 6b. The lattice fringe image was dominated by two spacings making 86° with each other and having periods 0.59 nm and 0.204 nm, not possible in the BCC cell. However, the Fourier transform implies close structural relationship of the Cr-rich phase large grains with the BCC phase formed at 450 °C ($a_{BCC} = 0.294 \pm 0.010$ nm). The structural relationship was illustrated by the directions corresponding to the BCC lattice in Figure 6b. Doubling the periodicity with respect to BCC phase was evident in the Fourier transform; however, based on the appearance of the 1/(4 × 0.294) nm^{-1} periodicity in the reciprocal lattice, (2 × 2 × 4)a_{BCC} size repeating units were deduced for the structure of this grain (Figure 6a). Additionally, a continuous/diffuse scattering of the $\frac{1}{2}\frac{1}{2}l$ reflections was observed parallel to the [001]*$_{BCC}$ direction, indicating non-periodicity along the crystallographic c axis (Figure 6b). In Figure 6c,d Fourier filtered images of the same area are shown.

In Figure 6c the periodic components corresponding to the BCC 001 and 110 distances are shown and all the other components were removed from the image by filtering. The image visualizes the periods, corresponding to $a_{BCC} = 0.294$ nm = d_{001} and at an angle of 90° to this, a period of spacing 0.208 nm ((110) planes in the BCC cell) representing the contribution of the BCC phase (Table 1) to the structure of these grains. Figure 6d, shows a Fourier filtered image made with the $\frac{1}{2}\frac{1}{2}l$ (BCC notation) diffuse scattering (continuous lines) in the insert of Figure 6b. The white arrows in the image indicate atomic planes parallel to 001 fringes seen in Figure 6c, however, at a rather random spacing (planar disorder). Nevertheless, the multiplies of the basic fringe spacing of the BCC lattice ($d_{002} = 0.147$ nm) in the [001] direction, i.e., the 0.294 nm and 0.588 nm (2 × a_{BCC}) still can be observed. Moreover, the [001] direction of the BCC lattice coincides with [001] of the (2 × 2 × 4) a_{BCC} size cell. So, the Cr-rich grains named A type in Figure 6a (and A in all other figures) can be considered a Cr-rich intermetallic phase with a unit cell 2 × 2 × 4 times the unit cell of the BCC phase, observed at and above 450 °C.

In Figure 7 the HREM analysis of the type "voids" (B in Figure 3a) was carried out. As one can see, these areas were really voids in the metallic film, however, they contained material in the form of an apparently amorphous layer and small, 5–10 nm in size crystalline particles. The amorphous layer could be considered to be mainly the supporting SiO$_x$ layer. The analysis of the FFT diffraction pattern results in an FCC phase with about 0.42 nm lattice parameter (Figure 7c).

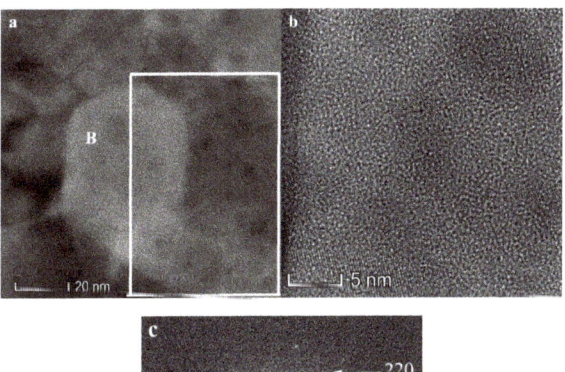

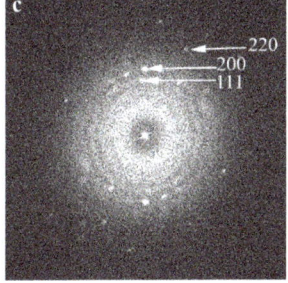

Figure 7. (**a**) Post-annealing high resolution image of the HEA (annealed at 700 °C) of an area marked B in Figure 3. White rectangle indicates the area of subsequent EDS mapping. (**b**) Magnified image of the area B in (**a**) showing crystalline particles. (**c**) Fast Fourier transform (FFT) diffraction of the region in (**b**) indexed according to an FCC oxide structure with lattice parameter a = 0.42 nm.

Local EDS measurement of the composition inside the "void" proves that these areas were dominated by Cr accompanied by only a little amount of the remaining HEA metal components (area #1 in Figure 8 and in Table 3). Areas #2 and #3 in Figure 8 and in Table 3 indicate compositional fluctuation on the tens of nm scale implying that the redistribution of components is going on. The composition measured in the whole area in Figure 8 demonstrates that the composition of the original HEA phase on average was basically preserved in the film. These results lead to the conclusion that the crystals, which remained in the voids have a possible composition $Cr_{4-x}M_xO_4$ (M can be Co, Fe, Ni or Cu) corresponding to four molecules in the FCC unit cell of either CrO [25] or a HEA-oxide [26] type phase.

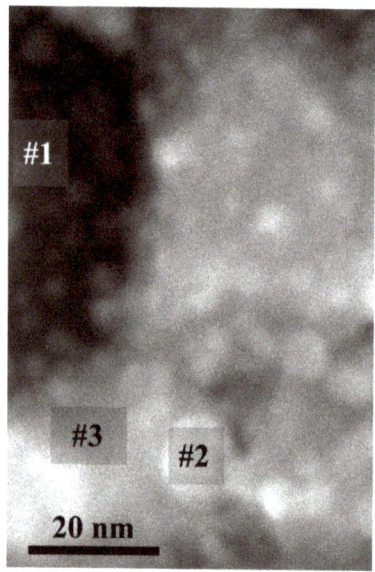

Figure 8. STEM HAADF image of the film annealed at 700 °C and the areas used for quantitative EDS analysis.

Table 3. Elemental composition measured in the areas marked in Figure 8. Analysis of spectrum: from a hole (Area#1), two small areas in the matrix (Areas #2 and #3) and the whole area of the STEM image in Figure 8.

Z	Element	Family	Atomic Fraction (%)			
			Area #1	Area #2	Area #3	The Whole Area
24	Cr	K	83.62	8.20	29.04	21.00
26	Fe	K	1.86	22.09	21.96	21.37
27	Co	K	1.56	18.84	19.61	19.95
28	Ni	K	0.66	20.93	17.28	19.44
29	Cu	K	12.29	29.94	12.10	18.24

4. Discussion

As we could clearly see from Figure 2, the changes of the structural transformations of the as-deposited CrFeCoNiCu HEA thin film could be divided into three temperature intervals. There were no changes below 400 °C, minor changes occur below 600 °C, and from this temperature rapid morphological transformations and separation of the components took place.

In the low temperature range below 400 °C at the present annealing periods no measurable changes occurred in the structure of the FCC CrFeCoNiCu HEA film.

During the annealing step at 450 °C a minor quantity of a new phase was observed with BCC lattice of a = 0.294 nm very close to that of Cr (a = 0.291 nm) and Fe (a = 0.288 nm). However, the identification of the grains of the BCC phase proved not to be straightforward. The BCC grains were located among the grains of the host FCC phase, having the same grain size. (Figure 2). This situation was preserved up to 550 °C. However, the nature of the FCC to BCC phase transformation was unclear. It could occur by nucleation and growth, probably accompanied by separation of components or by massive transformation without it. It could also happen by martensitic transformation, which according to [27], could occur by a simple deformation mechanism, leading to well defined crystallographic relationship of the starting FCC and the resulting BCC phases. In the case of the two diffusionless transformation mechanisms, i.e., massive and martensitic, the original random distribution of the components was preserved. In the case of diffusion, transformation implies chemical ordering of components (without their separation) in the originally single phase FCC HEA structure, containing the five constituting components in random distribution. Chemical ordering induces the formation of superstructures, which manifests itself in the appearance of extra reflections in the diffraction pattern with respect to the random FCC or BCC phases. As there were no extra (forbidden) reflections detected with respect to the HEA FCC and the BCC (consequently HEA BCC) phases up to 550 °C (Figure 2a) the preservation of the fully random distribution of the components was confirmed. This, together with the unchanged morphology supports that below 550 °C the kinetics of formation of the HEA BCC phase should take place by a diffusionless mechanism. Similar diffusionless transformation was reported for $Al_xCoFeNiCrCu$ bulk alloys [10] and attributed to the changing of Al content. In the present case the transformation in the CrFeCoNiCu HEA film occurred without any BCC forming additives and was triggered only by the increasing temperature. Summarizing, it can be inferred that the diffusionless formation mechanism of the HEA BCC phase is the first stage of transformations in the CrFeCoNiCu alloy films before the separation processes of components start and formation of further phases occurs.

Above 600 °C, (Figures 2 and 3), we could establish that besides the residual matrix, in which the FCC HEA grains represent the majority phase, other grains an order of magnitude larger in size exist. These larger grains, which mostly contain chromium, and indicate that separation of components occurred (Figure 4 and Table 2 as well as Figure 5), have superstructure reflections and also diffuse scattering in their FFT diffraction pattern. The superstructure reflections implying large unit cell parameters are integer multiples of the cell parameter of the HEA BCC phase. An example of such a Cr-rich grain with $2 \times 2 \times 4 a_{BCC}$ cell size is presented in Figure 6. We attributed the formation of the large ($2 \times 2 \times 4$ times the BCC) unit cell Cr-rich phase to the ordering processes during the ongoing separation of components in the HEA alloy.

As indicated by the diffuse scattering (Figure 6b) observed in the Fourier transforms of the Cr-rich grains, these grains exhibit an aperiodic structure as well, which is related to chemical disorder. The direction of diffuse scattering coincides with the [001]* reciprocal vector of the BCC phase and can be attributed to planar disorder in these crystals. The proposed planar disorder means, that the atomic planes of the large unit cell of the Cr-rich phase normal to the [001] direction do not have a real translational symmetry along [001] (Figure 6d) in the entire grain, but one or more of the five components concentrates on different (008) planes of the Cr-rich grain. However, other ordering patterns are also possible. This means, that the structure is regular for not more than a few unit cells (Figure 6d), then the ordering pattern changes. In some cases this can mean single atomic planes with different order/composition from their neighborhood resulting in "infinite" reciprocal rods, parallel to the [001]* reciprocal vectors as observed in Figure 6b in the form of the streaks. However, as individual reflections in the FFT diffraction pattern along [001]* containing the BCC reciprocal vectors remain visible, we suppose, that this kind of planar disorder is the property of the large cell, within which the composing former BCC cells (a = 0.294 nm) are preserved at least partly (Figure 6c). This explains the decrease of the number of BCC type reflections at temperatures above 650 °C as seen in Figures 2a and 3, because the reflections of the Cr-rich phase and the BCC phase coincide only partly. It also has to be noted that, according to our HRTEM observations, various pathways of separation and ordering may exist, the example presented is only one of them. This can account for the numerous scattered reflections in the diffraction pattern taken at 700 °C (Figure 3) practically without regular positioning, which hampers a general description of these structures. A possible phase, present in the annealed up to 700 °C films, is the sigma phase [28,29]. It is known that Cr is a strong sigma phase former below 700 °C, which can be deduced from the binary alloy phase diagrams of Cr-Co, Cr-Fe, or from Cr-Co-Fe ternary phase diagram [30]. The crystal structure of the sigma phase [29] belongs to the tetragonal space group with lattice parameters of a = 8.8 Å and c = 4.5 Å.

It was reported in the literature [10,21,22,31] that the nucleation pathway of Cr containing sigma phase prototype of Cr-Co and Cr-Fe alloys formed either through the BCC/B2 phase or directly from the FCC matrix. BCC/B2 phase prototypes can be transformed to sigma phase due to the high affinity of Cr toward Co, Cr, or Fe. In our case we believe that Cr–rich phase is developed from the HEA BCC phase observed in the 450–650 °C temperature interval and serves as a precursor of the sigma phase. This is supported also by the fact that the HEA BCC phase and the Cr-rich phase have a strict epitaxial relation.

From Table 1 we can see, that after the annealing at 700 °C the structure is probably composed of several phases not easily separable from each other due to the observed crystallographic similarities. The textured fraction of the FCC HEA phase is still the dominating one with a fraction of it transformed to the L12 phase. Having a closer look to the nature of the 0.248 nm ring (Figure 3) we could conclude that it has a mixed character. It has discrete diffraction spots, with a preferred orientation coinciding with that of the FCC HEA phase, which can be attributed to an ordered L12 phase. On the other hand, the 0.248 nm ring has a faint continuous ring as well, indicating a much smaller crystallite size. This ring we attributed to the FCC oxide phase, being a surface Cr-oxide with a = 0.42 nm as proven by HRTEM (Figure 8, Table 3). The most difficult task was the identification of the HEA BCC and the ordered Cr-rich phases in the diffraction pattern (Figure 3). HEA BCC rings completely overlapped with those of the L12 phase. Regarding the Cr-rich grains, the large periodicities corresponding to the $2 \times 2 \times 4a_{BCC}$ and similar unit cells were not detected in the in-situ SAED pattern (Figure 2a), maybe due to their low intensities (see also Figure 3). All the other reflections coincided with the FCC and L12 HEA phases, and could not be separated even on the basis of the continuous or spotty nature of the diffraction rings (Table 1). However, they were present in the FFT patterns of the high resolution images (Figure 6). Furthermore, post annealing elemental mapping of the samples (Figure 5) shows that small grains rich in either Cr or Cu are present in the structure indicating that the separation of components begins with Cr and Cu. So, the conclusion was that short time (5 min) 50 °C annealing steps result in a multiphase structure representing the initial stages of transformations, which were still far from equilibrium.

Summarizing the events above 600 °C, the HEA BCC grains served as precursors of the newly forming large Cr-rich grains and a transformation through ordering processes led to larger lattice periods in this phase. The grains of the Cr-rich phase were growing mainly laterally, remaining in the plane of the film (Figure 5 and areas 1 and 3 in Table 3). They were growing fast at the expense of the surrounding (still mainly nanocrystalline FCC HEA) matrix. During their growth, as it was related to the separation of the components from the surrounding matrix, the rate limiting kinetics must be volume diffusion. To get some insight into the kinetics of this process the activation energy of the growth rate of Cr-rich grains was estimated according to the procedure used in [32]. The calculated value of the activation energy for this process is 160 ± 30 kJ/mole and corresponded fairly well to that of bulk diffusion of Fe in Cu 187 kJ/mol in the temperature range between 650 and 1000 °C [33]. On this basis the rate limiting process in the high temperature range (T ≥ 600 °C) could be attributed to the lattice self-diffusion of components in the HEA alloy.

The strong affinity of Cr to oxygen resulting in a surface layer of Cr_3MO_4 (M can be: Fe, Co, Ni, or Cu) like composition in the void regions as well as over the whole film area (Figures 7 and 8) could be possibly responsible for an anticorrosive property of these alloys.

5. Conclusions

The CrFeCoNiCu HEA alloy films were stable in their FCC structure up to 400 °C.

At 450 °C a HEA BCC phase appeared in the film. This change was considered to occur by a diffusionless phase transformation. The structure, composition and grain size of the newly formed BCC grains corresponded to the HEA FCC grains.

The separation of the components began above 550 °C. The actual activation energy of this process was estimated to be around 160 kJ/mol. It must be related to lattice diffusion in the HEA alloy, which should be the rate limiting process during transformations above 550 °C. These transformations were:

- A new phase with a large unit cell, epitaxial to the already formed BCC phase started growing. The cell parameters correspond to $2 \times 2 \times 4$ units of the BCC cell. The crystals possessed a planar disorder of atomic planes in one of the <001> directions, though the BCC lattice was preserved as an internal skeleton of their structure.
- Formation of voids occurred in the metallic part of the film. These voids were, however, still containing a nano-crystalline phase of the composition close to CrO and having an FCC lattice of about 0.42 nm period. This must be the part of a surface chromium oxide layer, possibly contributing to corrosion resistance of these films.

Author Contributions: G.R. designed the project. M.A. and G.R. carried out the deposition and the in-situ TEM heating experiments. Post annealing analyses (HRTEM, elemental mapping) were done by V.K.K. All authors discussed the results and approved the final manuscript. All authors have read and agreed to the published version of the manuscript.

Funding: This study was funded by the Hungarian National Research, Development and Innovation Office through the OTKA NN112156 project.

Acknowledgments: The authors also acknowledge the VEKOP-2.3.3-15-2016-00002 project of the European Structural and Investment Funds. V.K.K. is indebted to the János Bolyai Fellowship of the Hungarian Academy of Sciences and the ÚNKP-19-4 New National Excellence Program of the Ministry for Innovation and Technology.

Conflicts of Interest: The authors declare no conflict of interest.

References

1. Cantor, B.; Chang, I.T.H.; Knight, P., Vincent, A.J.B Microstructural development in equiatomic multicomponent alloys. *Mater. Sci. Eng. A* **2004**, *375*, 213–218. [CrossRef]
2. Zhang, Y.; Zuo, T.T.; Tang, Z.; Gao, M.; Dahmen, K.A.; Liaw, P.K.; Lu, Z.P. Microstructures and properties of high entropy alloys. *Prog. Mater. Sci.* **2014**, *61*, 1–93. [CrossRef]
3. Gao, M.C. Progress in High-Entropy Alloys. *JOM* **2013**, *65*, 1749–1750. [CrossRef]
4. Lu, Y.P.; Gao, X.Z.; Jiang, L.; Chen, Z.N.; Wang, T.; Jie, J.C.; Li, T.J. Directly cast bulk eutectic and near-eutectic high entropy alloys with balanced strength and ductility in a wide temperature range. *Acta Mater.* **2017**, *124*, 143. [CrossRef]
5. Kao, Y.F.; Chen, T.J.; Chen, S.K.; Yeh, J.W. Microstructure and mechanical property of as-cast,-homogenized, and-deformed AlxCoCrFeNi ($0 \leq x \leq 2$) high-entropy alloys. *J. Alloy. Compd.* **2009**, *488*, 57. [CrossRef]
6. Miracle, D.B.; Senkov, O.N. A critical review of high entropy alloys and related concepts. *Acta Mater.* **2017**, *122*, 448–511. [CrossRef]
7. Xu, X.D.; Liu, P.; Guo, S.; Hirata, A.; Fujita, T.; Nieh, T.G.; Liub, C.T.; Chena, M.W. Nanoscale phase separation in a fcc-based CoCrCuFeNiAl0.5 high-entropy alloy. *Acta Mater.* **2015**, *84*, 145–152. [CrossRef]
8. Shun, T.T.; Hung, C.H.; Lee, C.F. Formation of ordered/disordered nanoparticles in FCC high entropy alloys. *J. Alloy. Compd.* **2010**, *493*, 105–109. [CrossRef]

9. Wang, J.; Niu, S.; Guo, T.; Kou, H.; Li, L. The FCC to BCC phase transformation kinetics in an Al0.5CoCrFeNi high entropy alloy. *J. Alloy. Compd.* **2017**, *710*, 144–150. [CrossRef]
10. Wang, W.R.; Wang, W.L.; Wang, S.C.; Tsai, Y.C.; Lai, C.H.; Yeh, J.W. Effects of Al addition on the microstructure and mechanical property of AlxCoCrFeNi high-entropy alloys. *Intermetallics* **2012**, *26*, 44–51. [CrossRef]
11. Pickering, E.J.; Muñoz-Moreno, R.; Stone, H.J.; Jones, N.J. Precipitation in the equiatomic high-entropy alloy CrMnFeCoNi. *Scr. Mater.* **2016**, *113*, 106–109. [CrossRef]
12. Otto, F.; Dlouhý, A.; Pradeep, K.G.; Kubenov, M.; Raabe, D.; Eggeler, G.; George, E.P. Decomposition of the single-phase high-entropy alloy CrMnFeCoNi after prolonged anneals at intermediate temperatures. *Acta Mater.* **2016**, *112*, 40–52. [CrossRef]
13. He, F.; Wang, Z.; Wu, Q.; Li, J.; Wang, J.; Liu, C.T. Phase separation of metastable CoCrFeNi high entropy alloy at intermediate temperatures. *Scr. Mater.* **2017**, *126*, 15–19. [CrossRef]
14. Zhang, Y.; Zhou, Y.J.; Lin, J.P.; Chen, G.L.; Liaw, P.K. Solid Solution Phase Formation Rules for Multi-Component Alloys. *Adv. Eng. Mater.* **2008**, *10*, 534–538. [CrossRef]
15. An, Z.; Jia, H.; Wu, Y.; Rack, P.D.; Patchen, A.D.; Liu, Y.; Ren, Y.; Li, N.; Liaw, P.K. Solid-solution CrCoCuFeNi high-entropy alloy thin films synthesized by sputter deposition. *Mater. Res. Lett.* **2015**, *3*, 203–209. [CrossRef]
16. Zhang, L.J.; Fan, J.T.; Liu, D.J.; Zhang, M.D.; Yu, P.F.; Jing, Q.; Ma, M.Z.; Liaw, P.K.; Li, G.; Liu, R.P. The microstructural evolution and hardness of the equiatomic CoCrCuFeNi high-entropy alloy in the semi-solid state. *J. Alloy. Compd.* **2018**, *745*, 75–83. [CrossRef]
17. Park, N.; Watanabe, I.; Terada, D.; Yokohama, Y.; Liaw, P.K.; Tsuji, N. Recrystallization Behavior of CoCrCuFeNi high-entropy alloy. *Metall. Mater. Trans. A* **2015**, *46*, 1481–1487. [CrossRef]
18. Oh, S.M.; Hong, S.I. Microstructural stability and mechanical properties of equiatomicCoCrCuFeNi, CrCuFeMnNi, CoCrCuFeMn alloys. *Mater. Chem. Phys.* **2018**, *210*, 120–125. [CrossRef]
19. Chen, T.K.; Shun, T.T.; Yeh, J.W.; Wong, M.S. Nanostructured nitride films of multi-element high-entropy alloys by reactive DC sputtering. *Surf. Coat. Technol.* **2004**, *188*, 193–200. [CrossRef]
20. Braeckman, B.R.; Misják, F.; Radnóczi, G.; Depla, D. The influence of Ge and in addition on the phase formation of CoCrCuFeNi high-entropy alloy thin films. *Thin Solid Film.* **2016**, *616*, 703–710. [CrossRef]
21. Dolique, V.; Thomann, A.L.; Brault, P.; Tessier, Y.; Gillon, P. Thermal stability of AlCoCrCuFeNi high entropy alloy thin films studied by in-situ XRD analysis. *Surf. Coat. Technol.* **2010**, *204*, 1989–1992. [CrossRef]
22. Rao, J.C.; Diao, H.Y.; Ocelík, V.; Vainchtein, D.; Zhang, C.; Kuo, C.; Tang, Z.; Guo, W.; Poplawsky, J.D.; Zhou, Y.; et al. Secondary phases in AlxCoCrFeNi high-entropy alloys: An in-situ TEM heating study and thermodynamic appraisal. *Acta Mater.* **2017**, *131*, 206–220. [CrossRef]
23. Nagase, T.; Rack, P.D.; Noh, J.H.; Egami, T. In-situ TEM observation of structural changes in nano-crystalline CoCrCuFeNi multicomponent high-entropy alloy (HEA) under fast electron irradiation by high voltage electron microscopy (HVEM). *Intermetallics* **2015**, *59*, 32–42. [CrossRef]
24. Zhang, Y.W.; Tunes, M.A.; Crespillo, M.L.; Zhang, F.X.; Boldman, W.L.; Rack, P.D.; Jiang, L.; Xu, C.; Greaves, G.; Donnelly, S.E.; et al. Thermal stability and irradiation response of nanocrystalline CoCrCuFeNi high-entropy alloy. *Nanotechnology* **2019**, *30*, 15. [CrossRef]
25. Lux, H.; Illmann, G. ZurKenntnis der Chrom (II) Salze und des Chrom (II) oxyds, II. *Chem. Ber.* **1959**, *92*, 9. [CrossRef]
26. Dedoncker, R.; Radnóczi, G.; Abadias, G.; Depla, D. Reactive sputter deposition of CoCrCuFeNi in oxygen/argon mixtures. *Surf. Coat. Technol.* **2019**, *378*, 124–362. [CrossRef]
27. Sandoval, L.; Urbassek, H.M.; Entel, P. The Bain versus Nishiyama–Wassermann path in the martensitic transformation of Fe. *New J. Phys.* **2009**, *11*, 103027. [CrossRef]
28. Jette, E.R.; Foote, F. The Fe-Cr alloy system. *Met. Alloy.* **1936**, *7*, 207.
29. Bergman, B.G.; Shoemaker, D.P. The space group of sigma-FeCr crystal structure. *J. Chem.Phys.* **1951**, *19*, 515. [CrossRef]
30. Okamoto, H.; Schlesinger, M.E.; Mueller, E.M. *Alloy Phase Diagram*; ASM International: Materials Park, OH, USA, 1992.
31. Shivam, V.; Basu, J.; Pandey, V.K.; Shadangi, Y.; Mukhopadhyay, N.K. Alloying behaviour, thermal stability and phase evolution in quinary AlCoCrFeNi high entropy alloy. *Adv. Powder Technol.* **2018**, *29*, 2221–2230. [CrossRef]

32. Nagy, K.H.; Misják, F. In-situ transmission electron microscopy study of thermal stability and carbide formation in amorphous Cu-Mn/C films for interconnect applications. *J. Phys. Chem. Solids* **2018**, *121*, 312–318. [CrossRef]
33. Prokoshkina, D.; Rodin, A.; Esin, V. About Fe Diffusion in Cu. *Defect Diffus. Forum* **2012**, *323*, 171–176. [CrossRef]

© 2020 by the authors. Licensee MDPI, Basel, Switzerland. This article is an open access article distributed under the terms and conditions of the Creative Commons Attribution (CC BY) license (http://creativecommons.org/licenses/by/4.0/).

Article

Micropillar Compression Study on the Deformation Behavior of Electrodeposited Ni–Mo Films

Jenő Gubicza [1],*, Garima Kapoor [1], Dávid Ugi [1], László Péter [2], János L. Lábár [1,3] and György Radnóczi [3]

1. Department of Materials Physics, Eötvös Loránd University, P.O.B.32, H-1518 Budapest, Hungary; garima_kpr_91@yahoo.com (G.K.); ugdtaat@caesar.elte.hu (D.U.); labar.janos@energia.mta.hu (J.L.L.)
2. Wigner Research Centre for Physics, Konkoly-Thege út 29-33, H-1121 Budapest, Hungary; peter.laszlo@wigner.mta.hu
3. Institute for Technical Physics and Materials Science, Centre for Energy Research, Konkoly-Thege út 29-33, H-1121 Budapest, Hungary; radnoczi.gyorgy@energia.mta.hu
* Correspondence: jeno.gubicza@ttk.elte.hu; Tel.: +36-1-3722876

Received: 26 January 2020; Accepted: 25 February 2020; Published: 27 February 2020

Abstract: The influence of Mo addition on the compression behavior of Ni films was studied by micropillar deformation tests. Thus, films with low (0.4 at.%) and high (5.3 at.%) Mo contents were processed by electrodeposition and tested by micropillar compression up to the plastic strain of about 0.26. The microstructures of the films before and after compression were studied by transmission electron microscopy. It was found that the as-deposited sample with high Mo concentration has a much lower grain size (~26 nm) than that for the layer with low Mo content (~240 nm). In addition, the density of lattice defects such as dislocations and twin faults was considerably higher for the specimen containing a larger amount of Mo. These differences resulted in a four-times higher yield strength for the latter sample. The Ni film with low Mo concentration showed a normal strain hardening while the sample having high Mo content exhibited a continuous softening after a short hardening period. The strain softening was attributed to detwinning during deformation.

Keywords: Ni–Mo films; micropillar compression; strain-softening; twins; detwinning

1. Introduction

Alloying with Mo is an effective way to tailor the physical properties of Ni. For instance, the Curie temperature decreases significantly with increasing Mo concentration in Ni [1]. For pure Ni, the temperature of transition from ferromagnetic to paramagnetic state is 354 °C which is reduced to about 60 °C when the Mo concentration increases to about 5 at.% [1]. On the other hand, the hardness and the wear resistance of Ni considerably increase with the addition of Mo, therefore Ni–Mo alloys are often used as hard coatings [2]. The improvement of hardness and wear resistance with grain refinement is a general phenomenon for Ni-based coatings, be achieved either with alloying or with incorporation of ceramic particles [3]. Alloying may result not only in the decrease of the grain size but also in the increase of the lattice defect density [4]. It has been shown that Mo addition enhances the density of twin faults [4] which improves the hardness since twin faults are obstacles against dislocation motion similar to grain boundaries [5]. In the case of alloying, the chemical nature of the added element also affects the mechanical properties, and this is why tungsten is a common candidate besides molybdenum for alloying nickel [6]. In dispersion-hardened coatings, the primary hardening factor is the grain refinement as a result of the particle incorporation, regardless of the properties of the particles incorporated [3].

As with their physical behaviour, Ni–Mo alloys also obtained attention due to their chemical properties. Ni–Mo is applied as a catalyst in hydrogen production either in the form of coating or powder since this material shows a high activity and a long-term stability in this process [7–9]. In bulk form, Ni–Mo is used as a substrate for epitaxially grown superconducting coatings as the sharp cube texture formed after rolling and subsequent annealing in Ni–Mo alloys is beneficial for the layer deposition [10]. Thus, Ni–Mo alloys have drawn significant attention from the scientific community due to their important applications in either bulk, powder or thin layer form.

Ni–Mo electroplated films can be produced by codeposition of Ni and Mo [11,12]. This is a typical induced codeposition process which means that the incorporation of Mo in the layer is induced by the deposition of Ni, i.e., pure Mo cannot be obtained by electroplating. The conditions of electrodeposition, such as bath composition, pH value, current density and stirring, influence strongly the concentration of the deposited Mo [13–15]. It was shown that the Mo content can reach 74 at.% in electrodeposited Ni films, but in this case the current efficiency became extremely low (about 1%) [15,16]. It has also been revealed that appropriate additives may facilitate the deposition of Mo in Ni [17].

Former studies revealed that the Mo content significantly influences the microstructure and hardness of Ni electrodeposits [4,18]. Namely, the grain size decreased while the density of lattice defects (e.g., dislocations and twin faults) increased with increasing Mo concentration. As a consequence, the hardness of the layers was considerably enhanced with the addition of Mo [18,19]. For instance, the hardness of a pure nanocrystalline Ni film was found to be about 4.3 GPa which increased to 5.5–6.0 GPa when 3–13 at.% Mo was codeposited with Ni [20]. It is noted that this hardness can be further enhanced with the application of annealing at 400–550 °C for 1 h [20,21]. This effect is referred to as anneal-hardening and may cause an increase of the hardness with a factor of two for electrodeposited Ni–Mo films. This hardening was explained by the segregation of Mo solutes to Ni grain boundaries which impeded both dislocation emission from grain boundaries and the grain boundary sliding during straining. Although the hardness of Ni–Mo films was studied extensively, the stress-strain response for these materials has not been studied yet.

In this paper, the deformation behavior of electrodeposited Ni–Mo films with lower (0.4 at.%) and higher (5.3 at.%) Mo contents was investigated by micropillar compression. This test has already been applied to the study of the mechanical properties of Ni, Ni-W and Ni-ceramic composite films [22–25]. At the same time, to the knowledge of the authors, this is the first micropillar compression on Ni–Mo films. It will be shown that not only the yield strength but also the strain-hardening behavior exhibits significant differences in the two layers. For the explanation of the different mechanical performances of the Ni films with low and high Mo contents, a detailed characterization of the microstructure was conducted before and after compression.

2. Materials and Methods

2.1. Film-Processing by Electrodeposition

Ni–Mo films were processed by electrodeposition at room temperature (RT) using a solution containing 0.52 mol/liter $NiSO_4$, 0.26 mol/liter sodium citrate, 0.1 g/liter sodium dodecylsulfate as wetting agent, and Na_2MoO_4 in varying concentration up to 6 mmol/liter. To minimize the impurity content of the films, a high-purity nickel sulfate salt with a Co concentration lower than 50 ppm was applied in the electroplating process. The pH of the bath was set as 6.1 ± 0.08 since this value yielded a very high Ni deposition efficiency (about 98%) [26]. Although saccharin is known as an efficient stress reliever for the deposition, it was not applied because the resulting sulfur content in the deposits may also impact the mechanical properties of the films. The current density was selected as -5.6 mA/cm^{-2}. This current density was about one order of magnitude lower than the values used commonly for the production of Ni–Mo films. The low current density yielded similarly high efficiency of deposition (96–98%) as obtained for pure Ni. Then, the Mo content in the films was tailored by changing the Mo concentration in the bath. Two films were deposited with low (0.4 ± 0.1 at.%) and high (5.3 ± 0.4 at.%) Mo contents. These values were determined by energy-dispersive X-ray spectroscopy (EDS) in an FEI

Quanta 3D scanning electron microscope (SEM, Thermo Fisher Scientific, Waltham, MA, USA). The electrodeposited samples with low and high Mo contents are denoted as LMo and HMo, respectively.

The deposition of the Ni–Mo films was carried out on a Cu substrate. First, the substrate was degreased and then placed horizontally at the bottom of the cell. A nickel wire spiral served as the counter electrode which was immersed into a frit-separated chamber of the cell in order to avoid the contamination of the deposit with the disintegrated grains of the anode. The deposition was stopped when the desired film thickness (about 20 µm) was achieved.

2.2. Microstructure Characterization of the As-Grown Film by Transmission Electron Microscopy

Transmission electron microscopy (TEM) was used for the determination of the average grain size in the as-deposited Ni–Mo films. The TEM samples were thinned by ion milling using liquid nitrogen cooling in order to avoid undesired annealing during thinning. In this procedure, GATAN G1 low temperature glue was used at 60 °C for fixing the sample in a Ti disk with a diameter of 3 mm. Then, the specimen was milled by Ar ions with the energy of 7 keV until perforation. The TEM experiments were carried out by a Philips CM20 electron microscope (Philips, Amsterdam, The Netherlands) operating at 200 keV. The mean grain size was determined as the average of the diameters of the grains identified in dark-field TEM images. About twenty grains were evaluated in this way for each film.

2.3. Characterization of the Crystallographic Texture of the Ni–Mo Films

The crystallographic texture of the films was characterized by the analysis of X-ray diffraction (XRD) pole figures which were measured by a Smartlab diffractometer made by Rigaku company, Japan using parallel-beam optics and Cu Kα radiation with the wavelength of 0.15418 nm. Before the pole figure measurements, diffraction patterns were taken by the Smartlab diffractometer using Bragg-Brentano geometry. The diffraction angles for reflections 111, 200 and 220 were determined for both Ni–Mo layers and these 2θ values were used in the pole figure measurements.

2.4. Micropillar Compression Test

The deformation behavior of the Ni–Mo films was studied by micropillar compression. Micropillars with square cross sections were fabricated by a focused ion beam (FIB) in the same SEM microscope as used in EDS experiments (see Section 2.1). The edge and the height of the pillars were 3 and 6 µm, respectively. The compression experiments were carried out by a home-made indenter device using a flat-ended cylindrical punch. The precision of the indentation depth and the load were ~1 nm and ~1 µN, respectively. In the present experiments, the maximum applied load was ~15 mN. The technical details of the indenter device can be found in reference [27]. To ensure the reproducibility of the compression data, three micropillars were fabricated and compressed for each film. SEM images were also taken on the pillars before and after deformation.

2.5. Characterization of the Microstructure of the Micropillars before and after Compression

The microstructure of the pillars before and after compression was studied by TEM and high-resolution TEM (HRTEM). First, thin sections parallel to the longitudinal axis of the pillars were cut using the FIB technique. Then, these foils were thinned by ion milling until perforation using Ar ions. The HRTEM structural characterization of the samples was carried out by a FEI Titan-Themis transmission electron microscope with a Cs corrected objective lens (point resolution is around 0.09 nm in HRTEM mode) operated at 200 kV.

3. Results

3.1. Microstructure of the As-Deposited Ni Films with Low and High Mo Contents

Figure 1 shows dark-field TEM images taken on the films LMo and HMo (lateral view). The average grain sizes determined from TEM images are ~240 and ~26 nm for samples LMo and HMo, respectively. Figure 1a also reveals that the large grains bordered by white dashed curves contain subgrains in the LMo film, appearing as bright and dark regions inside the grains. The size of these subgrains varies between 20 and 50 nm which is in good agreement with the diffraction domain size determined formerly by X-ray line profile analysis (XLPA) [4]. Namely, the average diffraction domain size was obtained as ~40 nm from fitting the experimental X-ray diffractogram using a theoretical pattern calculated for the description of the diffraction peak broadening caused by the ultrafine-grained microstructure [4]. For sample HMo, the X-ray diffraction domain size was ~47 nm which is slightly higher than the grain size determined by TEM. This difference can be explained by the many orders of magnitude larger volume studied by XLPA as compared to TEM. Nevertheless, the similar grain and diffraction domain sizes for layer HMo indicate that for this film the grains were not divided into subgrains.

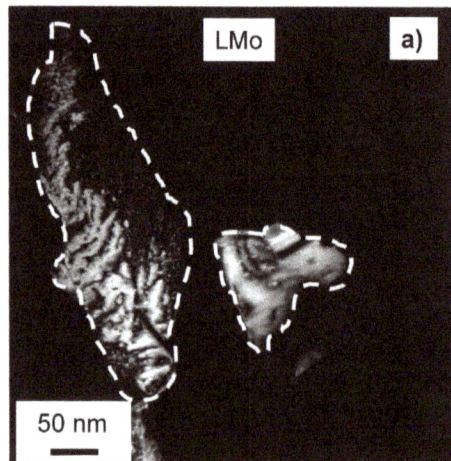
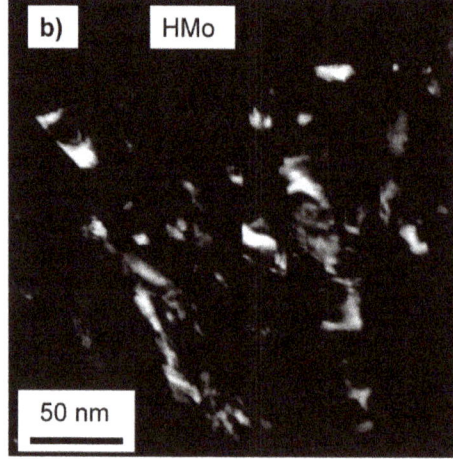

Figure 1. Dark-field transmission electron microscope (TEM) micrographs showing the grains in films with (**a**) low Mo content (LMo) and (**b**) high Mo content (HMo). In (**a**) the grains are bordered by white dashed curves for a better visibility.

The TEM image in Figure 2a shows that the nanograins in sample HMo contain twin faults. Former XLPA investigation revealed that the twin fault probability in layer HMo is as high as ~3.9% which corresponds to an average twin fault spacing of ~5 nm. This value is in accordance with the visual impression obtained from the HRTEM image in Figure 2b where some twin faults are indicated by white arrows. A magnified part of this HRTEM picture is Fourier-filtered in Figure 2c. In this image, only the (200) lattice fringes are shown. The yellow dashed lines indicate twin faults. The white arrow in Figure 2c marks the end of a twin lamella in the grain interior which usually comprises partial dislocations [28]. Sample LMo does not contain significant amount of twin faults as suggested by both TEM and XLPA since the twin fault probability determined by the latter method was under the detection limit (<0.1%) [4].

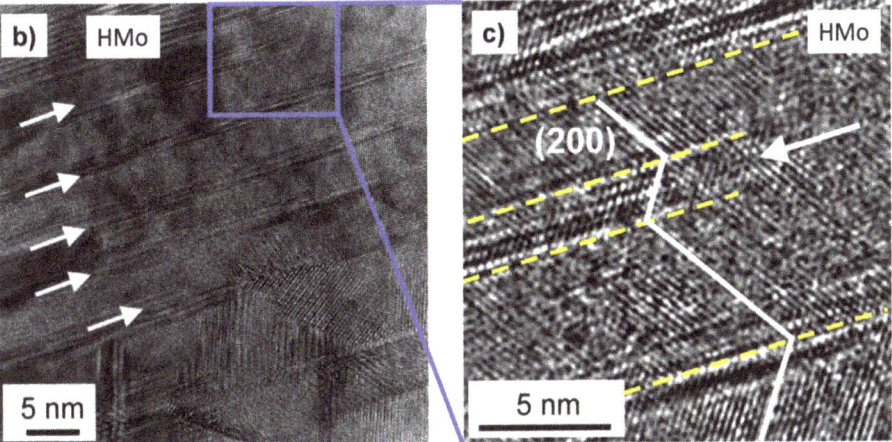

Figure 2. Bright-field TEM micrograph (**a**) and high-resolution TEM (HRTEM) image (**b**) taken on sample HMo. The white arrows indicate some twin faults. (**c**) shows a magnified and Fourier-filtered part of (**b**) (indicated by the blue frame). The dashed yellow lines mark twin faults. The white arrow in (**c**) indicates a twin lamella ending in the grain interior.

The crystallographic textures for samples LMo and HMo are characterized by the 111, 200 and 220 pole figures shown in Figure 3. The normal vector of the film surface is perpendicular to the plane of the pole figures. It is evident that the LMo film has a strong 200 texture parallel to the film normal. For sample HMo, no preferred orientation was detected.

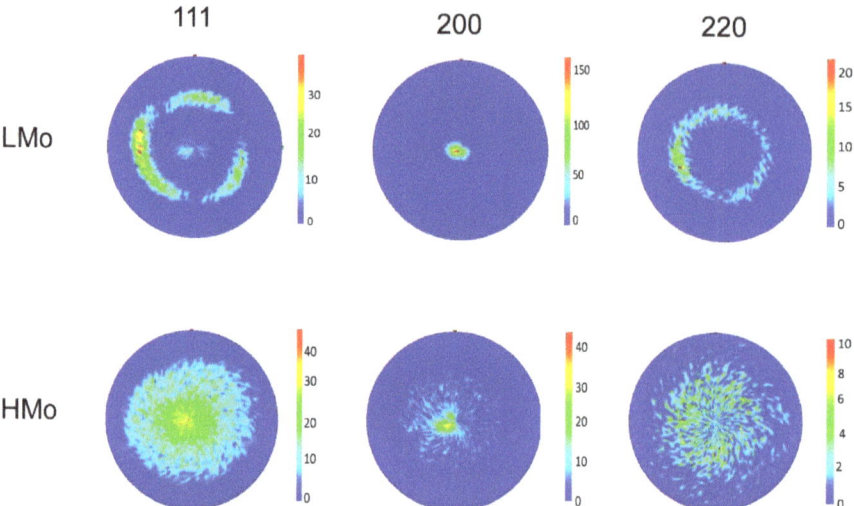

Figure 3. Pole figures for orientations 111, 200 and 220 as obtained by X-ray diffraction (XRD) for samples LMo and HMo. The normal vector of the film surface is perpendicular to the plane of the pole figures.

3.2. Compression Behavior of the Micropillars Fabricated from the Ni–Mo Films

As an example, Figure 4a,b show a micropillar fabricated from the film LMo before and after compression. The engineering stress versus plastic strain curves for samples LMo and HMo are plotted in Figure 5a,b, respectively. Very similar curves were obtained for other pillars manufactured from the same film. The engineering stress was obtained as the ratio of the applied force and the initial cross section of the pillars. The plastic portion of the engineering strain was calculated as follows. First, the engineering strain was determined as the ratio of the displacement and the initial pillar height (6 µm). Then, the elastic part of the strain was calculated as the ratio of the engineering stress and the elastic modulus. The latter quantity was determined as the slope of the initial linear part of the engineering stress-strain curve. Finally, the plastic strain was calculated as the difference between the total engineering strain and the elastic strain.

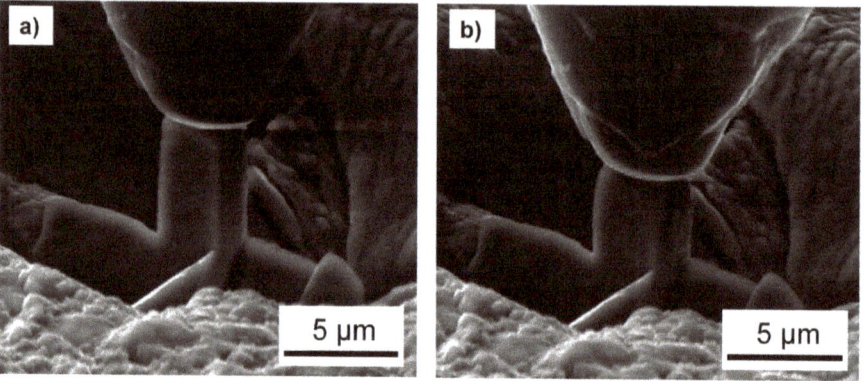

Figure 4. A micropillar under the indenter before (**a**) and after (**b**) compression for the film LMo.

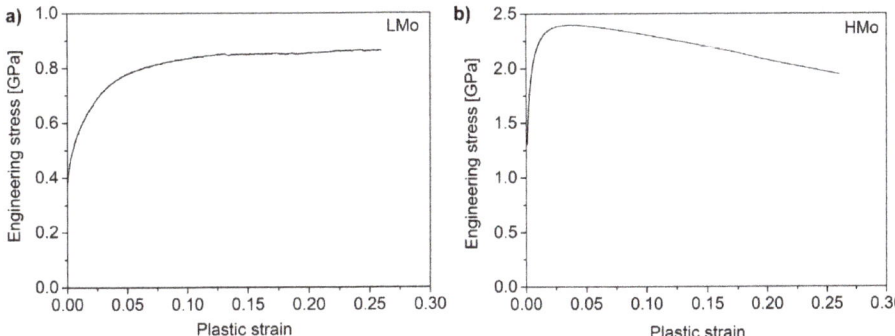

Figure 5. Engineering stress versus plastic strain obtained by micropillar compression for the films LMo (**a**) and HMo (**b**).

The yield strength values for samples LMo and HMo were obtained as 0.35 ± 0.05 and 1.3 ± 0.2 GPa, respectively. The sample LMo showed a monotonous hardening while the film HMo exhibited a strain-softening after an initial hardening stage. The maximum compressive stress values were 0.86 and 2.4 GPa for specimens LMo and HMo, respectively. These stresses were achieved at the plastic strains of 0.26 (at the end of the test) and 0.04 for the films LMo and HMo, respectively. The stress-strain behavior of sample LMo is not surprising; however, the strain-softening for film HMo is unusual. Therefore, the reason of this softening for sample HMo was studied by comparing the microstructures before and after micropillar deformation. These results are presented in the next section.

3.3. Changes of the Microstructure in the Ni Film with High Mo Content during Micropillar Compression

A TEM study was conducted on the microstructures of uncompressed and compressed micropillars manufactured from the film HMo. Figure 6 shows illustrative examples for the bright-field and the corresponding dark-field TEM images obtained before and after compression up to the plastic strain of 0.26. The average grain size determined from the images was about 24 nm for both the uncompressed and compressed micropillars, i.e., it remained unchanged during deformation. This value is practically the same as the grain size (~26 nm) obtained from the TEM images taken on the as-processed film HMo. It should be noted, however, that some larger grains with the size of about 100–200 nm were also found both before and after compression. As an example, a large grain in the compressed pillar is marked by a yellow ellipse in Figure 6d. This grain contains twin lamellas as revealed by the dark-field image in Figure 6d. It is noted that large grains were also observed in the uncompressed pillars. The numbers and the diameters of these grains were similar before and after compression. For example, long twin lamellas in a large grain can be seen at the bottom of Figure 6a taken on an uncompressed pillar. These large grains were formed in the nanocrystalline matrix during deposition and micropillar compression did not yield either their fragmentation or growth due to the relatively low plastic strain (about 0.26).

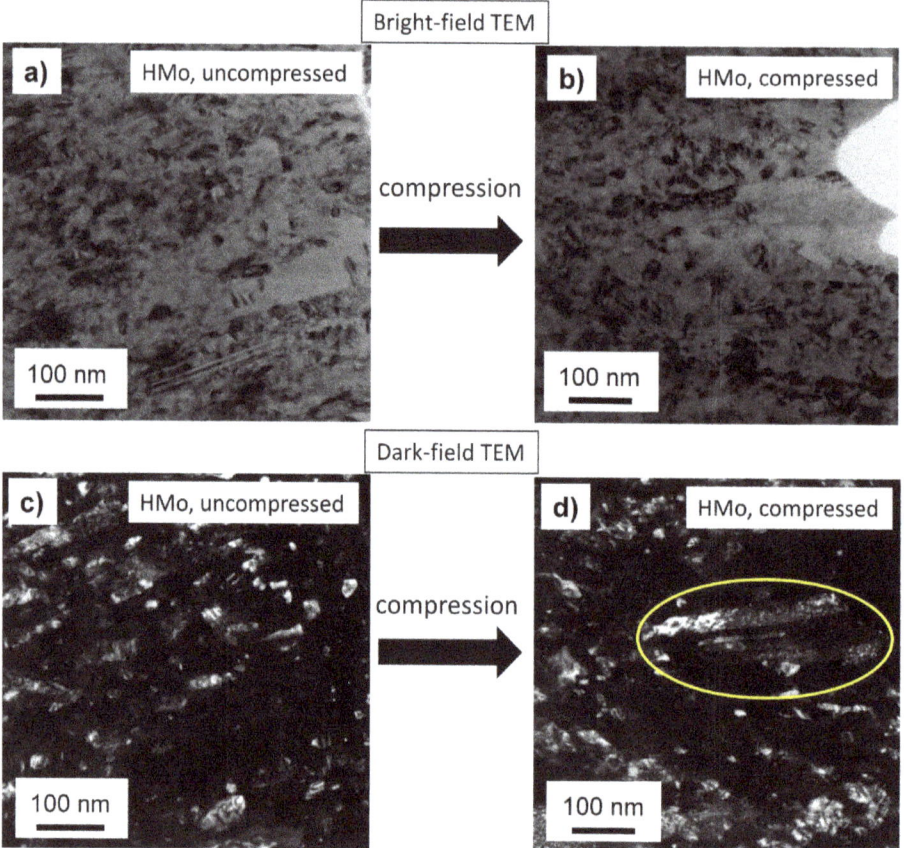

Figure 6. Bright-field (**a**,**b**) and the corresponding dark-field (**c**,**d**) TEM images taken on uncompressed (**a**,**c**) and compressed (**b**,**d**) micropillars for sample HMo.

The HRTEM images in Figure 7 show a high density of twin faults in both the uncompressed and compressed pillars for film HMo. The crystallographic direction <110> is lying perpendicular to the images. The twin boundaries are marked by white lines. Comparing Figure 7a,c, it seems that the twin fault density is lower for the compressed pillar than that in the undeformed state. Figure 7b shows a magnified part of Figure 7a. In this picture, bright and dark lattice fringes are visible parallel to the {111} planes and the periodicity of these fringes is three times larger than the lattice spacing for planes {111}. The Fourier transform of this HRTEM image can be seen in the lower right corner of Figure 7b which reveals that beside the fundamental fcc diffraction points additional spots appeared on the <111> lines of the reciprocal lattice, dividing the spacing between the fcc reciprocal lattice points into three. A former study explained these extra diffraction points by single and double diffraction from twins [29]. The different twin variants are indicated by numbers in Figure 7b. In some areas of Figure 7b, more than one twin variant exists overlapping each other in the TEM foil, which resulted in double diffraction of electrons.

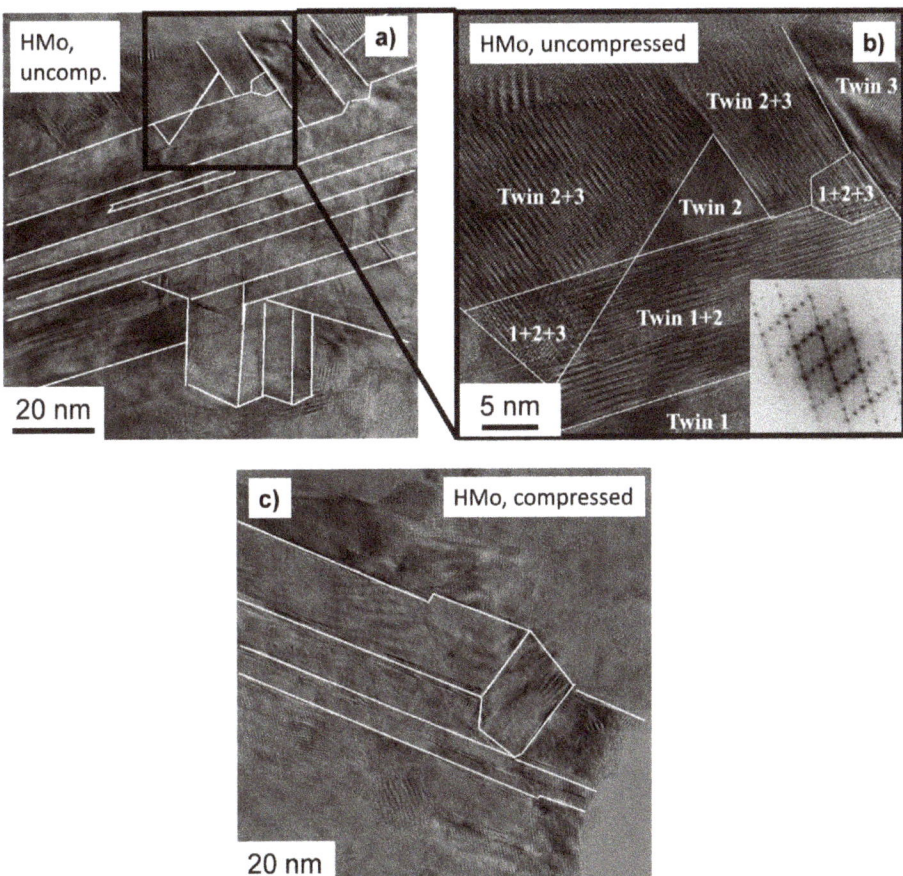

Figure 7. HRTEM images taken on uncompressed (**a,b**) and compressed (**c**) micropillars for the film HMo. The white lines indicate twin faults. (**b**) is a magnified part of (**a**) marked by the black frame. The diffraction pattern in (**b**) is obtained as the Fourier transform of the corresponding HRTEM image.

4. Discussion

The film HMo exhibited a much higher yield strength (~1.3 GPa) than that measured for sample LMo (~0.35 GPa). This difference can be explained by the combined hardening effect of the smaller grain size and the higher lattice defect (e.g., twin fault) density for the specimen HMo. Indeed, twin faults are similarly effective obstacles against dislocation motion such as the general grain boundaries [30,31]. Since the twin fault spacing in film HMo (about 5 nm) is smaller than the grain size (about 26 nm), the former value must be considered as the average distance between the dislocation glide obstacles. For sample LMo, either the grain size (~240 nm) or the crystallite size (~40 nm) is selected for the average obstacle spacing, it is much higher than the value determined for the film HMo, resulting in a softer yielding. In addition to the smaller obstacle spacing, the higher solute hardening for sample HMo may also contribute to the enhanced yield strength. Moreover, the different crystallographic textures in samples LMo and HMo also increased the difference between the yield strength values. Namely, film LMo exhibited a strong 200 texture and according to the Taylor model the Taylor factor for compression along direction <200> is about 2.4 which is lower than the value for an untextured fcc material (~3.06) [32]. The film HMo has no considerable texture (see Figure 3), therefore the Taylor factor for this material is surely higher than that for sample LMo during micropillar compression parallel to the film normal.

Figure 5 reveals that not only the yield strength but also the stress–strain behavior of the films LMo and HMo differ significantly. Namely, sample LMo exhibited strain hardening in the entire range of strain (up the strain of 0.26) while for film HMo softening was observed for the strains higher than ~0.04. Theoretically, softening can be caused by grain growth during plastic deformation. Indeed, former studies have shown that the high-pressure torsion (HPT) technique applied up to 30 turns on electrodeposited Ni–20% Fe caused an increase of the grain size from ~20 to ~50 nm at the equivalent strain of about 200 [33] and to ~115 nm when the equivalent strain increased to ~1300 [34]. This grain growth may be caused by a dynamic recrystallization of the as-deposited microstructure during deformation due to the high driving force owing to the large defect density and the small grain size. However, in sample HMo considerable grain growth did not occur during deformation, which can be attributed to the relatively low applied strain. Therefore, this effect is ruled out in the explanation of the softening observed for layer HMo.

The as-processed sample HMo contains a very high amount of grown-in twin faults, and detwinning during pillar compression might have occurred that could cause the observed softening. Detwinning is a result of the interaction between twin boundaries and gliding dislocations, yielding thinning or full disappearance of twin lamellae [35]. In the first step of detwinning, a gliding dislocation at the twin boundary dissociates into two partials. For a material with high stacking fault energy (SFE), these partials are sessile Shockley and glissile Frank dislocations [36]. The Shockley partial slips along the twin boundary, resulting in a thinning of the twin lamella with one {111} plane. This Shockley partial in the twin boundary is also called as twinning partial and it can also form if a dislocation is transmitted into the adjacent twin lamella [37]. The collective slip of twinning partials on successive glide planes parallel to the twin boundary can lead to a complete disappearance of a twinned region [28]. This effect has already been observed in electrodeposited Ni–20% Fe film processed by HPT [38]. Ni alloys have high SFE, therefore twinning is not a preferred mechanism of plastic deformation. At the same time, during the deposition of nanocrystalline pure Ni and Ni alloys many grown-in twin faults form as these have the lowest energy among the grain boundaries. Above the grain size of 20 nm, dislocation glide is an important deformation mechanism in Ni alloys [38], and therefore the interaction between moving dislocations and grown-in twin boundaries may cause detwinning. Then, the gradually decreasing twin fault density can result in a continuous softening during deformation as shown in Figure 5b. Indeed, the amount of twin faults in the film HMo seems to decrease during the present micropillar compression tests as suggested by the comparison of Figure 7a,c. The twin faults disappeared by detwinning were not replaced by new ones during compression as the probability of deformation twinning is very low in both studied Ni–Mo alloys as revealed in our former studies where the same compositions were deformed by HPT [39]. Up to the strain of about 1000, considerable twinning was not observed in bulk Ni samples with either ~0.3 or ~5 at.% of Mo. It should be noted that the reduction of the twin fault density during pillar compression of the film HMo is difficult to determine with good statistics from Figure 7 due to the small studied area (about 100 nm × 100 nm). However, the HRTEM images in Figure 7 suggest that in the investigated area the twin fault density decreased to about half during deformation.

Considerable reduction of grown-in lattice defects (e.g., dislocations and twin faults) during deformation of nanocrystalline fcc metals processed by bottom-up methods has already been observed in former studies [40]. This is a deformation-induced relaxation of nanostructures with extremely high density of growth defects. Due to the very small grain size and the extremely high defect density, nanocrystalline metals processed by bottom-up methods are very far from the equilibrium. However, this state can be frozen in the material as the annihilation of defects is strongly hindered kinetically by the impurities and the alloying elements. At the same time, plastic deformation of the as-processed samples causes a mechanical perturbation which can result in a shift of the material to a more equilibrium state by the annihilation of a portion of grown-in defects. For instance, a Ni–18 wt.% Fe alloy processed by pulsed electrodeposition was subjected to rolling at RT and liquid nitrogen temperature (LNT) up to the true strains between 0.4 and 0.6 [41]. This deformation resulted in a decrease of the twin fault probability from about 3.2% to 1.5–2.2%. In addition, the initial dislocation density was also reduced from 370×10^{14} m^{-2} to $(180-220) \times 10^{14}$ m^{-2} as revealed by XLPA. During plastic deformation of nanomaterials with the grain sizes higher than 10–20 nm, dislocations are emitted from the grain boundaries which slip across the host grain and are absorbed by the boundary at the opposite side of the grain. The interaction between the plasticity-induced and the growth dislocations can lead to their annihilation. The decrease of lattice defect density can result in softening of nanomaterials processed by bottom-up methods, as in the case of film HMo in the present study. It should be noted that softening in highly twinned microstructures was also observed for other fcc metals, such as Cu [42,43]. For copper, detwinning was detected not only during plastic deformation of nanotwinned films [43] but also for powders nanostructured by preliminary milling [44].

It is worth noting that during plastic deformation both defect formation and annihilation occur simultaneously and at high strains (>1) there is a dynamic equilibrium between these processes, resulting in a saturation of the values of the densities of lattice defects (e.g., dislocation and twin faults) [45]. If the density of grown-in defects in nanomaterials processed by bottom-up methods is higher than the saturation value achievable by severe plastic deformation, then deformation most probably results in a reduction of defect density. This was the case for sample HMo where the dislocation density in the electrodeposited film (~114×10^{14} m^{-2}) was much higher than the saturation value of ~60×10^{14} m^{-2} measured by XLPA on a sample processed from a coarse-grained material by HPT [39]. In addition, a high twin fault probability was detected in film HMo (~3.9%) while significant twin fault probability was not found in the sample processed till saturation by HPT. Therefore, defect density reduction was expected for layer HMo during micropillar compression. At the same time, in film LMo the density of growth dislocations was slightly lower (~23×10^{14} m^{-2}) than the saturation value achieved by HPT (~30×10^{14} m^{-2}). In addition, twins were not observed in either the electrodeposited or the HPT-processed LMo samples [39]. Thus, defect formation and a corresponding hardening were expected in this case which is in accordance with the present experimental observation. This research can be continued by studying the transition from strain hardening to softening as a function of Mo concentration in nanocrystalline Ni deposits.

5. Conclusions

The deformation behaviors of nanocrystalline Ni films deposited with low and high Mo contents were studied by micropillar compression test which was performed up to the plastic strain of 0.26. The following conclusions were drawn from the results:

- The film with high (5.3 at %) Mo concentration had a much larger yield strength (1.3 GPa) than the value obtained for low (0.4 at.%) Mo content (0.35 GPa). This difference can be attributed to the higher solute hardening, the much smaller grain size and the higher defect density in the former sample. In film HMo, nanotwins with an average spacing of ~5 nm were formed while considerable twinning was not observed in specimen LMo. In addition, a strong 200 texture was observed for film LMo while no considerable texture was detected in sample HMo, and this change also contributed to the higher yield strength of the latter specimen.
- The Ni film with low Mo concentration exhibited strain-hardening in the studied strain range, yielding a maximum compressive stress of 0.86 GPa. At the same time, layer HMo showed a fast hardening to the stress of 2.4 GPa which was followed by a continous softening between the strains of 0.04 and 0.26.
- The strain-softening for film HMo cannot be explained by grain coarsening since the average grain size remained about 26 nm during compression. On the other hand, a decrease of the twin density during compression was observed by comparing the TEM images taken on the pillars before and after deformation. This detwinning process caused the observed softening.

Author Contributions: Conceptualization, J.G.; methodology, L.P., J.L.L., G.R. and D.U.; formal analysis, G.K. and G.R.; investigation, G.K., J.L.L., G.R. and D.U.; resources, J.G., L.P., J.L.L. and G.R.; writing—original draft preparation, J.G.; writing—review and editing, G.K., D.U., L.P., J.L.L. and G.R.; visualization, J.G. and G.K.; supervision, J.G.; funding acquisition, J.G., L.P., J.L.L. and G.R. All authors have read and agreed to the published version of the manuscript.

Funding: This work was financed partly by the Ministry of Human Capacities of Hungary within the ELTE University Excellence program (1783-3/2018/FEKUTSRAT). This research was also funded by the Hungarian National Research, Development and Innovation Office through the OTKA NN112156 project. The VEKOP-2.3.3-15-2016-0000 project of the European Structural and Investment Funds is also acknowledged. This research was funded partly by EFOP-3.6.1-16-2016-00018 project: "Improving the role of research+development+innovation in the higher education through institutional developments assisting intelligent specialization in Sopron and Szombathely". The work was performed in the frame of Széchenyi 2020 program: "Innovative processing technologies, applications of energy engineering and implementation of wide range techniques for microstructure investigation".

Acknowledgments: The authors acknowledge the help of Éva Fekete in processing of the electrodeposited Ni–Mo films.

Conflicts of Interest: The authors declare no conflict of interest. The funders had no role in the design of the study; in the collection, analyses, or interpretation of data; in the writing of the manuscript, or in the decision to publish the results.

References

1. Karolus, M.; Lagiewka, E. The Structural Studies on nanocrystalline Ni-Mo alloys after annealing. In Proceedings of the XIX Conference of Applied Crystallography, Kraków, Poland, 1–4 September 2003; Morawiec, H., Stróz, D., Eds.; World Scientific: Singapore, 2004; pp. 337–341. [CrossRef]
2. Lehman, E.B.; Bigos, A.; Indyka, P.; Kot, M. Electrodeposition and Characterisation of Nanocrystalline Ni–Mo Coatings. *Surf. Coat. Technol.* **2011**, *211*, 67–71. [CrossRef]
3. Lekka, M. Electrochemical Deposition of Composite Coatings. In *Encyclopedia of Interfacial Chemistry*; Wandelt, K., Ed.; Elsevier: Amsterdam, The Netherlands, 2018; pp. 54–67. ISBN 9780128098943. [CrossRef]
4. Kapoor, G.; Péter, L.; Fekete, É.; Lábár, J.; Gubicza, J. The Influence of Mo Addition on the Microstructure and its Thermal Stability for Electrodeposited Ni Films. *Mater. Charact.* **2018**, *145*, 563–572. [CrossRef]
5. Kolonits, T.; Czigány, Z.; Péter, L.; Bakonyi, I.; Gubicza, J. Influence of Bath Additives on the Thermal Stability of the Nanostructure and Hardness of Ni Films Processed by Electrodeposition. *Coatings* **2019**, *9*, 644. [CrossRef]
6. Rupert, T.J.; Trelewicz, J.R.; Schuh, C.A. Grain boundary relaxation strengthening of nanocrystalline Ni-W alloys. *J. Mater. Res.* **2012**, *27*, 1285–1294. [CrossRef]
7. Brown, D.; Mahmood, M.; Man, M.; Turner, A. Preparation and Characterization of Low Overvoltage Transition Metal Alloy Electrocatalysts for Hydrogen Evolution in Alkaline Solutions. *Electrochim. Acta* **1984**, *29*, 1551–1556. [CrossRef]
8. Schulz, R.; Huot, J.Y.; Trudeau, M.L.; Bailey, L.D.; Yan, Z.H.; Jin, S.; Lamarre, A.; Ghali, E.; Neste, A.V. Nanocrystalline Ni-Mo alloys and their Application in Electrocatalysis. *J. Mater. Res.* **1994**, *9*, 2998–3008. [CrossRef]
9. McKone, J.; Sadtler, B.R.; Werlang, C.A.; Lewis, N.S.; Gray, H.B. Ni–Mo Nanopowders for Efficient Electrochemical Hydrogen Evolution. *ACS Catal.* **2013**, *3*, 166–169. [CrossRef]
10. Eickemeyer, J.; Selbmann, D.; Opitz, R.; Boer, B.; Holzapfel, B.; Schultz, L.; Miller, U. Nickel-Refractory Metal Substrate Tapes with High Cube Texture Stability. *Supercond. Sci. Technol.* **2001**, *14*, 152–159. [CrossRef]
11. Podlaha, E.; Landolt, D. An Experimental Investigation of Ni-Mo alloys. *J. Electrochem. Soc.* **1996**, *143*, 885–892. [CrossRef]
12. Podlaha, E.J.; Landolt, D. A Mathematical Model Describing the Electrodeposition of Ni-Mo Alloys. *J. Electrochem. Soc.* **1996**, *143*, 893–899. [CrossRef]
13. Chassaing, E.; Portail, N.; Levy, A.F.; Wang, G. Characterisation of Electrodeposited Nanocrystalline Ni-Mo Alloys. *J. Appl. Electrochem.* **2004**, *34*, 1085–1091. [CrossRef]
14. Bigos, A.; Beltowska-Lehman, E.; Kot, M. Studies on electrochemical deposition and physicochemical properties of nanocrystalline Ni-Mo alloys. *Surf. Coat. Technol.* **2017**, *317*, 103–109. [CrossRef]

15. Podlaha, E. Electrodeposition of High Mo Content Ni-Mo Alloys under Forced Convection. *J. Electrochem. Soc.* **1993**, *140*, L149–L151. [CrossRef]
16. Sun, S.; Podlaha, E. Electrodeposition of Mo-Rich, MoNi Alloys from an Aqueous Electrolyte. *J. Electrochem. Soc.* **2011**, *159*, D97–D102. [CrossRef]
17. Allahyarzadeh, M.H.; Roozbehani, B.; Ashrafi, A.; Shadizadeh, S.R.; Kheradmand, E. Electrodeposition of High Mo Content Amorphous/Nanocrystalline Ni-Mo Using Ionic Liquids as Additive. *ECS Trans.* **2012**, *41*, 11–28. [CrossRef]
18. Huang, P.C.; Hou, K.H.; Wang, G.L.; Chen, M.L.; Wang, J.R. Corrosion Resistance of the Ni-Mo Alloy Coatings Related to Coating's Electroplating Parameters. *Int. J. Electrochem. Sci.* **2015**, *10*, 4972–4984.
19. Wasekar, N.; Verulkar, S.; Vamsi, M.; Sundararajan, G. Influence of Molybdenum on the Mechanical Properties, Electrochemical Corrosion and Wear Behavior of Electrodeposited Ni-Mo alloy. *Surf. Coat. Technol.* **2019**, *370*, 298–310. [CrossRef]
20. Hu, J.; Shi, Y.; Sauvage, X.; Sha, G.; Lu, K. Grain Boundary Stability Governs Hardening and Softening in Extremely Fine Nanograined Metals. *Science* **2017**, *355*, 1292–1296. [CrossRef]
21. Zheng, X.; Hu, J.; Li, J.; Shi, Y. Achieving Ultrahigh Hardness in Electrodeposited Nanograined Ni-Based Binary Alloys. *Nanomaterials* **2019**, *9*, 546. [CrossRef]
22. Khalajhedayati, A.; Rupert, T.J. Emergence of localized plasticity and failure through shear banding during microcompression of a nanocrystalline alloy. *Acta Mater.* **2014**, *65*, 326–337. [CrossRef]
23. Mohanty, G.; Wheeler, J.M.; Raghavan, R.; Wehrs, J.; Hasegawa, M.; Mischler, S.; Philippe, L.; Michler, J. Elevated temperature, strain rate jump microcompression of nanocrystalline nickel. *Philos. Mag.* **2015**, *95*, 1878–1895. [CrossRef]
24. Mohanty, G.; Wehrs, J.; Boyce, B.L.; Taylor, A.; Hasegawa, M.; Philippe, L.; Michler, J. Room temperature stress relaxation in nanocrystalline Ni measured by micropillar compression and miniature tension. *J. Mater. Res.* **2016**, *31*, 1085–1095. [CrossRef]
25. Jarząbek, D.M.; Dziekoński, C.; Dera, W.; Chrzanowska, J.; Wojciechowski, T. Influence of Cu coating of SiC particles on mechanical properties of Ni/SiC co-electrodeposited composites. *Ceram. Int.* **2018**, *44*, 21750–21758. [CrossRef]
26. Mech, K.; Zabinski, P.; Mucha, M.; Kowalik, R. Electrodeposition of Catalytically Active Ni-Mo Alloys. *Arch. Metall. Mater.* **2013**, *58*, 227–229. [CrossRef]
27. Hegyi, Á.; Ispánovity, P.; Knapek, M.; Tüzes, D.; Máthis, K.; Chmelík, F.; Dankházi, Z.; Varga, G.; Groma, I. Micron-Scale Deformation: A Coupled in Situ Study of Strain Bursts and Acoustic Emission. *Microsc. Microanal.* **2017**, *23*, 1076–1081. [CrossRef]
28. Wang, J.; Li, N.; Anderoglu, O.; Zhang, X.; Misra, A.; Huang, J.; Hirth, J. Detwinning Mechanisms For Growth Twins In Face-Centered Cubic Metals. *Acta Mater.* **2010**, *58*, 2262–2270. [CrossRef]
29. Pashley, D.; Stowell, M. Electron Microscopy and Diffraction OF Twinned Structures in Evaporated Films of Gold. *Philos. Mag.* **1963**, *8*, 1605–1632. [CrossRef]
30. Lu, K.; Lu, L.; Suresh, S. Strengthening Materials by Engineering Coherent Internal Boundaries at the Nanoscale. *Science* **2009**, *324*, 349–352. [CrossRef]
31. Rath, B.B.; Imam, M.A.; Pande, M.A. Nucleation and Growth of Twin Interfaces in Fcc Metals and Alloys. *Mater. Phys. Mech.* **2000**, *1*, 61–66.
32. Clausen, B.; Lorentzen, T.; Leffers, T. Self-Consistent Modelling of the Plastic Deformation of F.C.C. Polycrystals and Its Implications for Diffraction Measurements of Internal Stresses. *Acta Mater.* **1998**, *46*, 3087–3098. [CrossRef]
33. Ni, S.; Wang, Y.B.; Liao, X.Z.; Alhajeri, S.N.; Li, H.Q.; Zhao, Y.H.; Lavernia, E.J.; Ringer, S.P.; Langdon, T.G.; Zhu, Y.T. Strain Hardening and Softening in a Nanocrystalline Ni–Fe Alloy Induced by Severe Plastic Deformation. *Mater. Sci. Eng. A* **2011**, *528*, 3398–3403. [CrossRef]
34. Ni, S.; Wang, Y.B.; Liao, X.Z.; Li, H.Q.; Figueiredo, R.B.; Ringer, S.P.; Langdon, T.G.; Zhu, Y.T. Effect of Grain Size on the Competition Between Twinning and Detwinning in Nanocrystalline Metals. *Phys. Rev. B* **2011**, *84*, 235401. [CrossRef]
35. An, X.; Ni, S.; Song, M.; Liao, X. Deformation Twinning and Detwinning in Face-Centered Cubic Metallic Materials. *Adv. Eng. Mater.* **2019**, *22*, 1900479. [CrossRef]
36. Mullner, P.; Solenthaler, C. On the Effect of Deformation Twinning on Defect Densities. *Mater. Sci. Eng. A* **1997**, *230*, 107–115. [CrossRef]

37. Jin, Z.H.; Gumbsch, P.; Albe, K.; Ma, E.; Lu, K.; Gleiter, H.; Hahn, H. Interactions between Non-Screw Lattice Dislocations and Coherent Twin Boundaries in Face-Centered Cubic Metals. *Acta Mater.* **2008**, *56*, 1126–1135. [CrossRef]
38. Ni, S.; Wang, Y.B.; Liao, X.; Figueiredo, R.B.; Li, H.Q.; Ringer, S.P.; Langdon, T.G.; Zhu, Y.T. The Effect of Dislocation Density on the Interactions between Dislocations and Twin Boundaries in Nanocrystalline Materials. *Acta Mater.* **2012**, *60*, 3181–3189. [CrossRef]
39. Kapoor, G.; Huang, Y.; Sarma, V.; Langdon, T.; Gubicza, J. Effect of Mo Addition on the Microstructure and Hardness of Ultrafine-Grained Ni Alloys Processed by a Combination of Cryorolling and High-Pressure Torsion. *Mater. Sci. Eng. A* **2017**, *688*, 92–100. [CrossRef]
40. Li, L.; Ungar, T.; Wang, Y.D.; Fan, G.J.; Yang, Y.L.; Jia, N.; Ren, Y.; Tichy, G.; Lendvai, J.; Choo, H.; et al. Simultaneous Reductions of Dislocation and Twin Densities with Grain Growth During Cold Rolling in a Nanocrystalline Ni–Fe alloy. *Scr. Mater.* **2009**, *60*, 317–320. [CrossRef]
41. Li, L.; Ungár, T.; Wang, Y.D.; Morris, J.R.; Tichy, G.; Lendvai, J.; Yang, Y.L.; Ren, Y.; Choo, H.; Liaw, P.K. Microstructure Evolution During Cold Rolling in a Nanocrystalline Ni–Fe Alloy Determined by Synchrotron X-Ray Diffraction. *Acta Mater.* **2009**, *57*, 4988–5000. [CrossRef]
42. Han, J.; Sun, J.; Han, Y.; Zhu, H.; Fang, L. Strengthening versus Softening of Nanotwinned Copper Depending on Prestress and Twin Spacing. *Metals* **2018**, *8*, 344. [CrossRef]
43. Lu, L.; Chen, X.; Huang, X.; Lu, K. Revealing the Maximum Strength in Nanotwinned Copper. *Science* **2009**, *323*, 607–610. [CrossRef] [PubMed]
44. Wen, H.; Zhao, Y.; Li, Y.; Ertorer, O.; Nesterov, K.M.; Islamgaliev, R.K.; Valiev, R.Z.; Lavernia, E.J. High-Pressure Torsion-Induced Grain Growth and Detwinning in cryomilled Cu powders. *Philos. Mag.* **2010**, *90*, 4541–4550. [CrossRef]
45. Gubicza, J. *Defect Structure and Properties of Nanomaterials*, 2nd ed.; Woodhead Publishing: Duxford, UK, 2017; ISBN 9780081019184.

© 2020 by the authors. Licensee MDPI, Basel, Switzerland. This article is an open access article distributed under the terms and conditions of the Creative Commons Attribution (CC BY) license (http://creativecommons.org/licenses/by/4.0/).

Article

Determination of the Complex Dielectric Function of Ion-Implanted Amorphous Germanium by Spectroscopic Ellipsometry

Tivadar Lohner [1,†], Edit Szilágyi [2,†], Zsolt Zolnai [1,†], Attila Németh [2,†], Zsolt Fogarassy [1,†], Levente Illés [1,†], Endre Kótai [2,†], Peter Petrik [1,†,*] and Miklós Fried [1,3]

1. Institute of Technical Physics and Materials Science, Centre for Energy Research, Konkoly Thege Miklós Rd. 29-33, H-1121 Budapest, Hungary; lohner@mfa.kfki.hu (T.L.); zolnai.zsolt@energia.mta.hu (Z.Z.); fogarassy.zsolt@energia.mta.hu (Z.F.); illes.levente@energia.mta.hu (L.I.); fried.miklos@energia.mta.hu (M.F.)
2. Institute for Particle and Nuclear Physics, Wigner Research Centre for Physics, Konkoly Thege Miklós Rd. 29-33, H-1121 Budapest, Hungary; szilagyi.edit@wigner.hu (E.S.); nemeth.attila@wigner.mta.hu (A.N.); kotai.endre@wigner.mta.hu (E.K.)
3. Institute of Microelectronics and Technology, Óbuda University, P.O. Box 112, 1431 Budapest, Hungary
* Correspondence: petrik@mfa.kfki.hu
† These authors contributed equally to this work.

Received: 9 April 2020; Accepted: 11 May 2020; Published: 16 May 2020

Abstract: Accurate reference dielectric functions play an important role in the research and development of optical materials. Libraries of such data are required in many applications in which amorphous semiconductors are gaining increasing interest, such as in integrated optics, optoelectronics or photovoltaics. The preparation of materials of high optical quality in a reproducible way is crucial in device fabrication. In this work, amorphous Ge (a-Ge) was created in single-crystalline Ge by ion implantation. It was shown that high optical density is available when implanting low-mass Al ions using a dual-energy approach. The optical properties were measured by multiple angle of incidence spectroscopic ellipsometry identifying the Cody-Lorentz dispersion model as the most suitable, that was capable of describing the dielectric function by a few parameters in the wavelength range from 210 to 1690 nm. The results of the optical measurements were consistent with the high material quality revealed by complementary Rutherford backscattering spectrometry and cross-sectional electron microscopy measurements, including the agreement of the layer thickness within experimental uncertainty.

Keywords: germanium; optical properties; dielectric function; thin film characterization; semiconductor; spectroscopic ellipsometry; optical dispersion; Tauc-Lorentz model; Cody-Lorentz model

1. Introduction

Accurate and reliable optical data of materials are scarce in the literature, although they are of key importance for the modeling of coatings, as well as optical or structural materials [1,2]. Ge, its alloys, as well as many other crystalline and amorphous semiconductors, especially Si, Ge and their compounds are used as detectors [3], Bragg reflectors [4], photodiodes [5], materials of controlled optical properties (especially in the infrared wavelength range [6]), band gap [7] and refractive index [1] engineering.

The optical properties and the thickness of thin film structures can be derived from (Ψ,Δ) values measured by spectroscopic ellipsometry (SE), where Ψ and Δ describe the relative amplitude and relative phase change, respectively [8]. SE is the primary tool to determine the optical properties and structure of materials [9], in many cases utilizing the in situ capabilities [10,11]. Concerning amorphous Ge (a-Ge) films, papers dealing with the optical and structural characterization of evaporated Ge layers can be found in the literature [12–16], and only a few papers discuss the optical and structural characterization of a-Ge layers obtained by low energy (0.5–1.0 keV) ion bombardment [17,18]. Aspnes and Studna irradiated single-crystalline Ge (c-Ge) surfaces using Ne and Ar ions with 1 keV energy. They performed SE measurements and determined the dielectric function of the ion bombardment-amorphized Ge (ia-Ge) layers. They obtained a 9 nm thick ia-Ge layer for 1-keV Ne bombardment and determined its dielectric function [17]. This layer thickness can be considered as ultra-thin and even an atomically thin transition "layer" between the c-Ge and ia-Ge region can cause more than 10% uncertainty. We measured a thick layer and the weighted uncertainty caused by the transition layer is low.

To describe the optical properties on an amorphous material as a function of photon energy or wavelength the Tauc-Lorentz (TL) or Cody-Lorentz (CL) dispersion models are frequently used. The TL model was developed by Jellison and Modine [19] to provide a dispersion equation for a material that only absorbs light above the material bandgap. The CL analytical model elaborated by Ferlauto et al. was designed to model optical properties of amorphous materials [20].

In this work, we used ion implantation to create a high-density void-free amorphous material for reference database purposes. We showed that by the proper choice of the implantation parameters (element, multiple energies, angle, etc.), a layer is formed that is comparable with the highest qualities found in the literature in terms of optical density. Additional to the optical references of amorphous Ge currently available, we provide an analytical model that well described the dispersion in a broad wavelength range. The results of the optical characterizations were verified by complementary methods including Rutherford backscattering spectrometry combined with channeling (RBS/C) and cross-sectional transmission electron microscopy (XTEM).

2. Experimental Details

A Ge wafer from Umicore (orientation of (100), resistivity of approx. 0.4 Ωcm, CAS Nr. 7440-56-4) was cleaned in diluted HF (CAS Nr. 7664-39-3) and rinsed in deionized (DI) water. After cutting it into small rectangular pieces, the samples were rinsed again in DI and dried in N gas. To produce a homogeneous amorphous layer from the surface to the buried crystalline-amorphous interface the amorphized layer was created by two step amorphization via ion implantation at room temperature (first step 120-keV Al^+ (CAS Nr. 7429-90-5) at a fluence of 1×10^{16} atoms/cm^2; second step: 300-keV Al^+ 1×10^{16} atoms/cm^2) using a heavy-ion cascade implanter (Figure 1a). To avoid the channeling effect during implantation, the sample was tilted by 7° with respect to the ion beam (Figure 1b). Although after the first implantation the Al ions may partially channelled even at tilt 7° [21], the amorphous layer formed by the first ion implantation ensured that the Al ions in the second ion implantation step couldn't practically get channelled in the sample. The reason for selecting a relatively light mass projectile (Al) was to avoid void formation in case of implantation of heavy mass ions [22].

The damage level caused by ion implantation can be characterised by the displacements per atoms (DPA), i.e., the number of times that an atom is displaced for a given fluence. Figure 1c shows the Al and DPA distribution determined by the simulation software of Stopping and Range of Ions in Matter (SRIM) [23]. SRIM calculates the number of displacements per one implanted ion and per unit depth as a function of depth in the irradiated sample. Considering the implanted fluence (the number of implanted ions per unit area), the value of DPA can be determined for the applied fluence. At depths where the calculated DPA reaches the threshold value (0.3 DPA in our case [24]), the target is supposed to turn to an amorphous phase from crystalline. In our case the DPA value exceeds the threshold from the sample surface down to a depth of 630 nm. Therefore, based on the SRIM simulations, we suppose the position of the a-Ge/c-Ge interface to be at the depth of 630 nm.

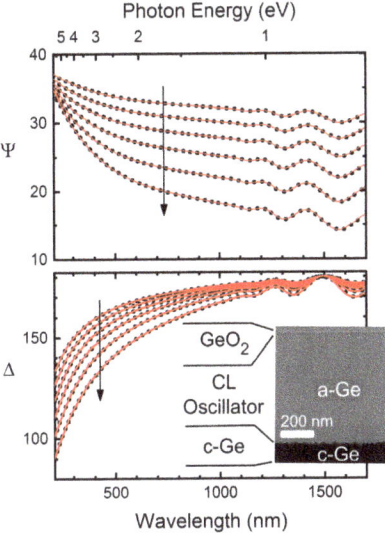

Figure 5. Measured and generated ellipsometric Ψ and Δ spectra for the Al-implanted Ge sample. The two-layer optical model and the corresponding XTEM micrograph revealing a completely amorphized layer with a thickness of 681 nm is shown in the inset. The fitted thicknesses of the surface oxide (GeO$_2$) and the amorphous Ge (CL oscillator) layers are 1.74 ± 0.01 and 678.9 ± 0.1 nm, respectively. Note that the top part of the XTEM image is the glue used for the sample preparation. The 1.7-nm oxide itself is not visible. The arrows show the direction of increasing angles of incidence from 53° in steps of 3°.

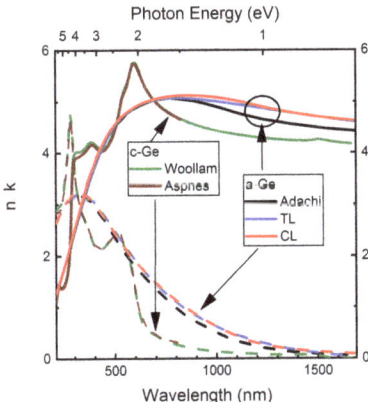

Figure 6. n (solid lines) and k (dashed lines) spectra given by the evaluation of the measured SE data using the CL and TL dispersions. For comparison, the data of c-Ge [28,30] and evaporated a-Ge [15] (Adachi) are also presented.

The thickness values of the ia-Ge layer determined by SE and the theoretical thickness of damaged region estimated by SRIM agree with the thickness value determined by RBS/channeling technique within the experimental uncertainty of RBS. However, SE gives a somewhat larger thickness value compared to RBS; probably SE is more sensitive to damage than the RBS/channeling technique.

The images obtained by XTEM investigation are shown in the insets of Figures 3 and 5. The HRTEM image shown in Figure 4 and its fast Fourier transrom (FFT) obtained by XTEM method (as well as the insets of Figures 3 and 5) show a completely amorphized germanium layer. This result justifies the appropriate choice of Al for the ion implantation, because a high quality, void-free, dense and completely amorphous Ge layer was formed. The density of the a-Ge layer found in our study is even higher than that was reported in Ref. [15]. This result is also reflected in our n and k values which are slightly higher than that shown in Ref. [15], especially at higher wavelengths.

4. Conclusions

The complex dielectric function of ia-Ge produced by ion implantation was determined by SE in the wavelength range from 210 to 1690 nm. It was found that the CL dispersion relation is more appropriate for the evaluation of the SE measurements on ia-Ge than the TL model. The thickness values yielded by the TL and by the CL type SE evaluations are close to the thickness value deduced from the ion beam analytical measurements and XTEM investigation. The obtained dielectric function spectra are in good agreement with those measured by Adachi et al. [15], providing a solid and reliable basis of further in situ investigations of amorphization processes in Ge.

Author Contributions: T.L. performed the ellipsometry measurements, the evaluations and wrote the manuscript; A.N. performed the ion irradiation; E.S., Z.Z., and E.K. measured and evaluated the RBS/C spectra; Z.F. and L.I. measured by electron microscopy; P.P. wrote the manuscript and coordinated the work; M.F. coordinated the work. All authors have read and agreed to the published version of the manuscript.

Funding: This research was funded by the National Development Agency grant OTKA K131515, VEKOP-2.3.3-15-2016-00002 and grant NKFIH-OTKA 129009.

Conflicts of Interest: The authors declare no conflict of interest. The funders had no role in the design of the study; in the collection, analyses, or interpretation of data; in the writing of the manuscript, or in the decision to publish the results.

Abbreviations

The following abbreviations are used in this manuscript:

CL model	Cody-Lorentz model
DI	Deionized Water
DPA	Displacements per Atoms
FFT	Fast Fourier Transform
MSE	Mean Square Error
RBS	Rutherford Backscattering Spectrometry
SE	Spectroscopic Ellipsometry
SRIM	Stopping and Range of Atoms in Matter
TL model	Tauc-Lorentz model
XTEM	Cross-sectional Transmission Electron Microscopy

References

1. Lohner, T.; Kalas, B.; Petrik, P.; Zolnai, Z.; Serényi, M.; Sáfrán, G. Refractive Index Variation of Magnetron-Sputtered a-Si$_{1-x}$Ge$_x$ by 'One-Sample Concept' Combinatory. *Appl. Sci.* **2018**, *8*, 826. [CrossRef]
2. Kalas, B.; Zolnai, Z.; Safran, G.; Serenyi, M.; Agocs, E.; Lohner, T.; Nemeth, A.; Fried, M.; Petrik, P. Micro-combinatorial sampling of the optical properties of hydrogenated amorphous Si$_{1-x}$Ge$_x$ for the entire range of compositions towards a database for optoelectronics. *Sci. Rep.* **2020**, submitted.
3. Cosentino, S.; Miritello, M.; Crupi, I.; Nicotra, G.; Simone, F.; Spinella, C.; Terrasi, A.; Mirabella, S. Room-temperature efficient light detection by amorphous Ge quantum wells. *Nanoscale Res. Lett.* **2013**, *8*, 128. [CrossRef] [PubMed]
4. Leem, J.W.; Yu, J.S. Design and fabrication of amorphous germanium thin film-based single-material distributed Bragg reflectors operating near 2.2 μm for long wavelength applications. *JOSA B* **2013**, *30*, 838–842. [CrossRef]
5. Colace, L.; Balbi, M.; Masini, G.; Assanto, G.; Luan, H.C.; Kimerling, L.C. Ge on Si p-i-n photodiodes operating at 10Gbit/s. *Appl. Phys. Lett.* **2006**, *88*, 101111, doi:10.1063/1.2182110. [CrossRef]

6. Carletti, L.; Sinobad, M.; Ma, P.; Yu, Y.; Allioux, D.; Orobtchouk, R.; Brun, M.; Ortiz, S.; Labeye, P.; Hartmann, J.M.; et al. Mid-infrared nonlinear optical response of Si-Ge waveguides with ultra-short optical pulses. *Opt. Express* **2015**, *23*, 32202. [CrossRef]
7. Bean, J. Silicon-based semiconductor heterostructures: Column IV bandgap engineering. *Proc. IEEE* **1992**, *80*, 571–587. [CrossRef]
8. Fujiwara, H. *Spectroscopic Ellipsometry: Principles and Applications*; John Wiley & Sons Ltd.: Chichester, UK, 2007.
9. Castriota, M.; Politano, G.G.; Vena, C.; De Santo, M.P.; Desiderio, G.; Davoli, M.; Cazzanelli, E.; Versace, C. Variable Angle Spectroscopic Ellipsometry investigation of CVD-grown monolayer graphene. *Appl. Surf. Sci.* **2019**, *467–468*, 213–220. [CrossRef]
10. Nemeth, A.; Kozma, P.; Hülber, T.; Kurunczi, S.; Horvath, R.; Petrik, P.; Muskotál, A.; Vonderviszt, F.; Hős, C.; Fried, M.; et al. In Situ Spectroscopic Ellipsometry Study of Protein Immobilization on Different Substrates Using Liquid Cells. *Sens. Lett.* **2010**, *8*, 730–735. [CrossRef]
11. Fricke, L.; Böntgen, T.; Lorbeer, J.; Bundesmann, C.; Schmidt-Grund, R.; Grundmann, M. An extended Drude model for the in-situ spectroscopic ellipsometry analysis of ZnO thin layers and surface modifications. *Thin Solid Films* **2014**, *571*, 437–441. [CrossRef]
12. Al-Mahasneh, A.; Al Attar, H.; Shahin, I. Spectroscopic ellipsometry of single and multilayer amorphous germanium/aluminum thin film systems. *Opt. Commun.* **2003**, *220*, 129–135. [CrossRef]
13. Beaglehole, D.; Zavetova, M. The fundamental absorption of amorphous Ge, Si and GeSi alloys. *J. Non-Cryst. Solids* **1970**, *4*, 272–278. [CrossRef]
14. Rafla-Yuan, H.; Rancourt, J.; Cumbo, M. Ellipsometric study of thermally evaporated germanium thin film. *Appl. Opt.* **1997**, *36*, 6360–6363. [CrossRef] [PubMed]
15. Adachi, S. *Optical Constants of Crystalline and Amorphous Semiconductors: Numerical Data and Graphical Information*; Springer: New York, NY, USA, 1999.
16. Wei, P.; Xu, Y.; Nagata, S.; Narumi, K.; Naramoto, H. Structure and optical properties of germanium implanted with carbon ions. *Nucl. Instrum. Methods Phys. Res. Sect. B: Beam Interact. Mater. Atoms* **2003**, *206*, 233–236. [CrossRef]
17. Aspnes, D.; Studna, A. An investigation of ion-bombarded and annealed 111 surfaces of Ge by spectroscopic ellipsometry. *Surf. Sci.* **1980**, *96*, 294–306. [CrossRef]
18. Dekker, J.; Zandvliet, H.; van Silfhout, A. Low energy ion bombardment on c-Ge surfaces. *Vacuum* **1990**, *41*, 1690–1691. [CrossRef]
19. Jellison, G.E.; Modine, F.A. Erratum: "Parameterization of the optical functions of amorphous materials in the interband region" [Appl. Phys. Lett. 69 , 371 (1996)]. *Appl. Phys. Lett.* **1996**, *69*, 2137–2137. [CrossRef]
20. Ferlauto, A.S.; Ferreira, G.M.; Pearce, J.M.; Wronski, C.R.; Collins, R.W.; Deng, X.; Ganguly, G. Analytical model for the optical functions of amorphous semiconductors from the near-infrared to ultraviolet: Applications in thin film photovoltaics. *J. Appl. Phys.* **2002**, *92*, 2424–2436. [CrossRef]
21. Nordlund, K.; Djurabekova, F.; Hobler, G. Large fraction of crystal directions leads to ion channeling. *Phys. Rev. B* **2016**, *94*, 214109. [CrossRef]
22. Kaiser, R.; Koffel, S.; Pichler, P.; Bauer, A.; Amon, B.; Claverie, A.; Benassayag, G.; Scheiblin, P.; Frey, L.; Ryssel, H. Honeycomb voids due to ion implantation in germanium. *Thin Solid Films* **2010**, *518*, 2323–2325. [CrossRef]
23. Ziegler, J.F.; Ziegler, M.; Biersack, J. SRIM – The stopping and range of ions in matter (2010). *Nucl. Instrum. Methods Phys. Res. Sect. B: Beam Interact. Mater. Atoms* **2010**, *268*, 1818–1823. [CrossRef]
24. Birtcher, R. Energy dependence of amorphization of Ge by Kr ions. *MRS Online Proc. Libr. Arch.* **1993**, *279*, 129–134. [CrossRef]
25. Pászti, F.; Manuaba, A.; Hajdu, C.; Melo, A.; Da Silva, M. Current measurement on MeV energy ion beams. *Nucl. Instrum. Methods Phys. Res. B* **1990**, *47*, 187–192. [CrossRef]
26. Kótai, E. Computer methods for analysis and simulation of RBS and ERDA spectra. *Nucl. Inst. Methods Phys. Res. B* **1994**, *85*, 588–596. [CrossRef]
27. Woollam Co., Inc. Available online: http://www.jawoollam.com (accessed on 15 April 2020).
28. (Ge Tabulated from UNL (Multiple Data Sets Fit)—Data in the Woollam Software Library). Available online: /WVASE32new/mat/Ge.mat (accessed on 13 April 2020).

29. Fried, M.; Lohner, T.; Aarnink, W.; Hanekamp, L.; Van Silfhout, A. Determination of complex dielectric functions of ion implanted and implanted-annealed amorphous silicon by spectroscopic ellipsometry. *J. Appl. Phys.* **1992**, *71*, 5260–5262. [CrossRef]
30. Aspnes, D.E.; Studna, A.A. Dielectric functions and optical parameters of Si, Ge, GaP, GaAs, GaSb, InP, InAs, and InSb from 1.5 to 6.0 eV. *Phys. Rev. B* **1983**, *27*, 985–1009. [CrossRef]

Sample Availability: The dielectric function data determined for the amorphous Ge layer created by implantation of Al ions can be downloaded from https://seafile.it.energia.mta.hu/f/32291c1f099541ba855d/?dl=1.

© 2020 by the authors. Licensee MDPI, Basel, Switzerland. This article is an open access article distributed under the terms and conditions of the Creative Commons Attribution (CC BY) license (http://creativecommons.org/licenses/by/4.0/).

Article

X-ray Diffraction Investigation of Stainless Steel—Nitrogen Thin Films Deposited Using Reactive Sputter Deposition

Faisal I. Alresheedi [1,2] and James E. Krzanowski [3,*]

1. Physics Department, University of New Hampshire, 9 Library Way, Durham, NH 03824, USA; fia1@wildcats.unh.edu
2. Department of Physics, College of Science, Qassim University, Buraidah 51452, Saudi Arabia
3. Mechanical Engineering Department and Materials Science Program, University of New Hampshire, 33 Academic Way, Durham, NH 03824, USA
* Correspondence: James.Krzanowski@unh.edu; Tel.: +1-603-862-2315

Received: 15 September 2020; Accepted: 12 October 2020; Published: 15 October 2020

Abstract: An X-ray diffraction investigation was carried out on nitrogen-containing 304 stainless steel thin films deposited by reactive rf magnetron sputtering over a range of substrate temperature and bias levels. The resulting films contained between ~28 and 32 at.% nitrogen. X-ray analysis was carried out using both the standard Bragg-Brentano method as well as area-detector diffractometry analysis. The extent of the diffraction anomaly ((002) peak shift) was determined using a calculated parameter, denoted R_B, which is based on the (111) and (002) peak positions. The normal value for R_B for FCC-based structures is 0.75 but increases as the (002) peak is anomalously displaced closer to the (111) peak. In this study, the R_B values for the deposited films were found to increase with substrate bias but decrease with substrate temperature (but still always >0.75). Using area detector diffractometry, we were able to measure d_{111}/d_{002} values for similarly oriented grains within the films, and using these values calculate c/a ratios based on a tetragonal-structure model. These results allowed prediction of the (002)/(200) peak split for tetragonal structures. Despite predicting a reasonably accessible split (~0.6°–2.9°–2θ), no peak splitting observed, negating the tetragonal-structure hypothesis. Based on the effects of film bias/temperature on R_B values, a defect-based hypothesis is more viable as an explanation for the diffraction anomaly.

Keywords: sputter deposition; thin films; X-ray diffraction; expanded austenite

1. Introduction

Since the initial discovery of the S-phase by Zhang and Bell [1] and Ichii et al. [2], understanding the structural nature of this phase and the anomalous shift of the (200)/(400) diffraction peaks has been a challenging problem. The S-phase (also termed "expanded austenite") was discovered as a result of research aimed at creating a nitrogen-enriched surface layer on stainless steels for improved wear resistance. In the mid-to-late 1990's much of the research was centered on the investigation of low-temperature plasma nitriding methods [3–10] and the development of surface hardening methods via a combination of nitrogen implantation and diffusion. These processes generally are carried out within the temperature range of 250–400 °C; at temperatures above 400, CrN forms, depleting the matrix of Cr and reducing corrosion resistance, while below 250 °C nitrogen diffusion is too slow to form a surface layer of significant depth. Numerous studies on plasma nitriding methods for stainless steels have been reported including ion beam implantation [3–5] and the plasma immersion implantation method [5–10]. Structural characterization of treated surfaces revealed the formation of the S-phase, and a significant degree of surface hardening was observed along with substantial reductions in wear rates.

In addition to plasma nitriding methods, the S-phase can be produced by sputter deposition from stainless steel targets in a nitrogen-containing atmosphere [11–16]. Saker et al. [11] and Bourjot et al. [12] reported deposition of "stainless-steel nitrogen" coatings deposited by triode reactive magnetron sputtering from 310 stainless targets. A nitrogen content of up to 42% was obtained and the S-phase was confirmed by X-ray diffraction. The microhardness was measured and the maximum was reported as 15 GPa at a nitrogen concentration of about 15%. Shedden at al. [13] deposited coatings from 316 stainless steel using magnetron sputtering and a substrate temperature of 350 °C. They found the nitrogen content in the films increased with the proportion of N_2 in the sputtering gas, and reached a maximum of about 40%. The films had a very strong <100> fiber texture, although the fiber axis tilted away from the substrate normal at the highest N_2 flow rates. In addition, they examined the formation of energetic neutrals during sputtering and showed the yield of nitrogen energetic neutrals was much greater than that for argon. Therefore, as nitrogen content of the sputter gas increased, the burial of nitrogen within the growing films also increased, indicating enrichment with nitrogen was a primarily dynamic phenomenon.

The nitrogen content in the films described above were all sub-stoichiometric, i.e., with N/Me < 1, and contained up to 40% nitrogen. However, higher nitrogen concentrations have been obtained by increasing the percentage of N_2 in the sputtering gas during deposition. Kappaganthu and Sun [15] deposited films from a 316L target in an Ar+N_2 mixture with nitrogen contents ranging from 0 to 75% (at a constant sputtering pressure of 0.67 Pa.) The nitrogen content in the films increased with the -percent N_2 in the gas mixture and reached a maximum of 50% film nitrogen at N_2 content of 50% and higher. For film nitrogen concentrations between 35% and 45%, the (200) peak position anomaly was observed; however, for films with 50% nitrogen the d-spacings measured were all consistent with a single lattice parameter. The MeN (Me = Fe, Cr, Ni and Mo) phase was proposed to have a zinc-blende type structure. Kappaganthu and Sun [16] also examined the effect of substrate temperature and found that deposition at 300 °C promoted single-phase S-phase formation, but at 400 °C some CrN formation was observed.

The (200)/(400) peak position anomaly (characterized by observations where the position of the (200)/(400) peaks are inconsistent with the remaining peaks in the XRD patterns) has puzzled investigators for over 30 years, and there is still no consensus on the structural features of the S-phase that cause this peak shift. However, four main explanations have been proposed: (1) the S-phase is not a single phase but consists of multiple phases; (2) the structure is not FCC but rather (slightly) distorted into a tetragonal, monoclinic or other structure; (3) the anomaly is due to a high density of stacking faults; and (4) the anomaly results from a very large anisotropy in elastic constants. Early investigations by Marchev et al. [17,18] led to the claim that the S-phase has a tetragonal structure, and in fact, they re-named the structure as the "m-phase" due to its similarity to bct martensite. In this case, the X-ray diffraction patterns should show split (200)/(002) peaks. However, no such peak split was observed, but its absence was attributed to the pronounced crystallographic texture (in this case a (111) orientation) in the samples. Bacci et al. [19] also claimed that an fct-structured S-phase provided a reasonable fit to their diffraction data, but the presence of the S-phase in the form of a compositionally-varying diffusion layer, well as iron nitride phases, complicated the analysis.

To further examine the possibilities of non-cubic structures, Fewell et al. [20] conducted TEM and XRD studies of plasma-nitrided AISI316 steel. In addition to the traditional Bragg-Brentano XRD method, they used a second beam angle (non-zero ψ) in order to measure a set of d-spacings for the same grain orientations (relative to the surface). They found no evidence for multiple phases and noted that the diffraction data again showed only expanded (200) and (400) planes. Attempts were made to rationalize this in terms of tetragonal, monoclinic and triclinic structures. The triclinic gave the best fit to the diffraction data; however, due to the broadening of the S-phase peaks, a definitive conclusion could not be made. Fewell and Priest [21] then examined the S-phase using synchrotron radiation, allowing them to conduct higher-order diffractometry and d-spacing measurements out to the (622) planes. They presented an extensive analysis of numerous non-cubic structures, but found that none of them worked well in matching the higher-order reflections.

Numerous investigations have pointed to stacking faults (on {111} planes of the fcc structure) or other defects as an explanation for the diffraction anomaly [22–24]. The basis of this approach is the theoretical analysis presented by Warren [25] who determined the effect of stacking faults on the peak positions. The peak shifts were given in terms of the stacking fault density α ($1/\alpha$ is the number of planes between faults) and (hkl)-dependent constants. In most cases, the value of α is determined based on the $\Delta 2\theta$ calculated from the peak shift in the (200) reflection. For example, Blawert et al. [22] found $\alpha = 0.167$ for their nitrogen expanded austenite samples; Christiansen and Somers [23] used $\alpha \sim 0.03$ to obtain results consistent with their data; and while Xu et al. [24] did not give a specific value for α, they noted it should be dependent on nitrogen content. In order to unambiguously test the stacking fault hypothesis, it would be necessary to independently measure the stacking fault density α, and compare the calculated $\Delta 2\theta$ values with the observed shifts. However, this does not appear to have been done in any of the above studies, although stacking faults have been observed in several TEM studies. Xu et al. [26] and Stroz and Psoda [27] both examined the microstructure of plasma nitrided samples and observed stacking faults in the S-phase; the high-resolution image in the latter study showed stacking fault bundles with $\alpha \sim 0.1$. Nonetheless, they proposed the peak shift was due to a slight rhombohedral distortion in the lattice. The stacking fault explanation has been criticized in a number of papers [26,28] due to the fact that Warren's model becomes inaccurate at high values of α. A more detailed analysis of stacking fault effects was carried out by Velterop et al. [29]. However, the general effects described by Warren still hold, and for the (200) reflections only slight changes to the calculated $\Delta 2\theta$ values appear to be necessary. Another problem with the stacking fault theory is that for the (400) reflections the peaks should shift in the opposite direction (to higher angles). However, careful measurements, such as those made by Fewell and Priest, show a decrease in the (400) position which is similar in magnitude to the (200) shift. Therefore, the stacking fault hypothesis does not seem consistent with much of the data.

The final explanation for the diffraction anomaly is the elastic anisotropy hypothesis. Grigull and Parascandola [30] carried out a residual stress analysis for the S-phase layer to determine the strain perpendicular and parallel to the surface. The residual stress increased dramatically with nitrogen content in the layer, and at 23% N the (compressive) stress was 2.5–3 GPa. Abrasonis et al. [31] found the strain in (100) oriented grains (relative to the surface) to be twice that of (111)-oriented grains. However, they used elastic constants for nitrogen-free austenitic stainless steel, since the elastic constants of the S-phase are not known. Nonetheless, they suggested that the combination residual stress and stacking fault effects could explain the diffraction anomaly.

The possibility of ordering of nitrogen atoms on the interstitial sublattice has been considered and potential evidence for such ordering was recently presented by Brink et al. [32]. The presence of such ordering would require indexing of diffraction patterns based on a larger unit cell, and this unit cell could have a distorted (non-cubic) shape. Ordering may also influence the distribution of metal atoms on the metal sublattice as shown in a recent EXAFS (Extended X-ray Absorption Fine Structure) study [33]. Another recent study by Czerwiec et al. [34], where Mössbauer spectroscopy was used to examine the detailed atomic structure in annealed 316L nitride samples, proposed that the structure consisted of two different environments: a one which was supersaturated with nitrogen, and another consisting of a martensitic environment without nitrogen.

In summary, the structure of the S-phase still remains controversial as none of the four hypotheses appears adequate to explain all of the observed results. A recent article by Christiansen et al. [35] concluded that stacking faults, composition gradients and residual stress gradients provided the best explanations the observed X-ray diffraction pattern anomalies in plasma treated bulk stainless steel samples.

In this work, we analyze films sputter-deposited from 304 stainless steel targets in a nitrogen-containing atmosphere and characterize these samples using X-ray diffraction methods.

The novelty of this work stems from the following observation: when the Bragg-Brentano method is employed, measured d-spacings for (111) and (200) planes are made from grains of different orientations. In this case, stress and elastic anisotropy effects can impact measured values. Ideally, measurements of both d-spacings should be made from grains of similar orientations. This was done by Fewell et al. [20] for select orientations. However, by using area detector diffactrometry a continuous range of orientations can be examined, which will be done here. We can then compare these results with those obtained using the Bragg-Bretano method. In addition, the possibility of a tetragonal-based structure will be examined.

In addition to insight gained from the use of area-detector diffractometry, the use of sputter-deposited samples deposited with variations in temperature and bias allow further understanding into the effects of composition and defect content on the structure of the S-phase [36]. For this purpose, samples will first be characterized using the Bragg-Bretano method, where the extent of the diffraction anomaly will be evaluated using the following term:

$$R_B = \frac{\sin^2 \theta_{111}}{\sin^2 \theta_{200}} \tag{1}$$

where θ_{111} and θ_{200} are the peak positions obtained (by definition) from an X-ray diffraction scan carried out using the standard Bragg-Bretano configuration, i.e., with $\psi = 0$. The normal value of R_B for an FCC structure is 0.75, and a value of $R_B > 0.75$ indicates that the sample has the S-phase structure.

Following this analysis we will consider the S-phase structure as nominally FCC (rocksalt structure) with a slight deviation along one cube direction resulting in a tetragonal structure. The assumption of tetragonality is taken to allow the parameter c/a to be calculated based on the equation presented by Fewell and Priest [21], given by:

$$\frac{c}{a} = \left(\frac{1}{2} \left[\frac{3a^2_{002}}{a^2_{111}} - 1 \right] \right)^{1/2} \tag{2}$$

Alternatively, this equation can be written in terms of d-spacings:

$$\frac{c}{a} = \sqrt{\frac{2d^2_{002}}{d^2_{111}} - \frac{1}{2}} \tag{3}$$

Which is more amenable to direct calculation from X-ray diffraction data and makes no presumptions about the relationships between a_{hkl} and d_{hkl} values. Here we have assumed that the c-axis corresponds to the [001] direction and $c/a > 1$. Using this equation, measurements of the (111) and (002) peak positions allow for determination of the c/a ratio. This will be done using d_{200} and d_{111} values from the same grain, or grains of the same orientation.

The method devised for this purpose is illustrated in Figure 1, which shows a schematic diagram of a cross-section of a film with a typical columnar structure. We assume a fiber texture for the grain structure, and define the variable φ as the angle of tilt of the [001] direction away from the nominal surface normal (substrate plane). Three cases are shown in the diagram. For grain 1, the grain orientation is [002] so that $\varphi = 0°$. The value of d_{002} for this grain can be determined by conducting an XRD scan with $\psi = 0$. However, to determine d_{111} the scan needs to be run with $\psi = 54.74°$. For grain 2, $\varphi = 54.74°$ but the value of d_{111} is determined with $\psi = 0°$, and to find d_{002} we set $\psi = 54.74°$. For grain 3, we examine an intermediate orientation, in this case a grain with a [411] orientation. This gives $\varphi = 19.5°$, necessitating the use of $\psi = 19.5°$ to find d_{002} and $\psi = 35.27°$ to find d_{111}.

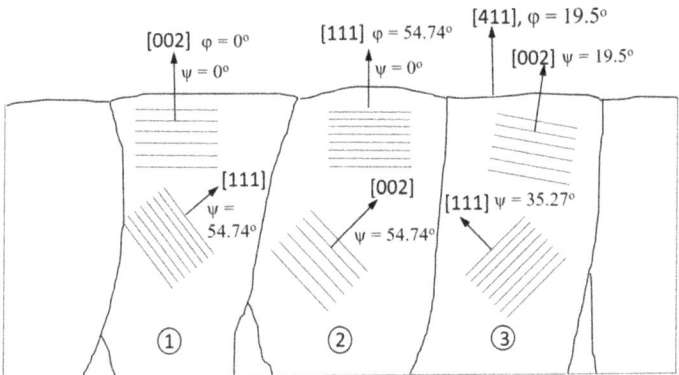

Figure 1. Schematic diagram of typical grain orientations showing the definitions of φ and ψ relative to the crystal structures and sample surface. The φ angle is the angle of tilt away from the [002] direction in the crystal, whereas ψ is the common diffraction vector, and is perpendicular to the surface in a Bragg-Brentano diffraction experiment.

In general, we can write:
For d_{002}: φ = ψ
For d_{111}: φ = 54.74 − ψ

Therefore, if we can find d as a function ψ for the (111) and (200) reflections, the above equations can be used to find d_{hkl} vs. φ and then determine c/a using Equation (3). Using this procedure, we find c/a as a function of φ. If there is no elastic anisotropy present, or in the absence of stress, c/a should be constant with φ.

2. Materials and Methods

Films were deposited using rf-magnetron sputtering in a turbo-molecular pumped high vacuum system. The base pressure was 2×10^{-6} Torr (0.266 mPa) and the total gas pressure during sputtering was 5 mTorr (0.67 Pa). Si (100) wafers were used as substrates and 304 stainless steel was used for the target. The substrate-to-target distance was 12 cm, with the sputter sources at an angle of 14° from the substrate normal direction. The sputter power density was 7.5 W/cm^2, and the rf frequency was 13.56 MHz. In order to improve adhesion of films to the Si substrates, a metallic stainless steel film was first deposited in Ar at −50 V bias to a thickness of 50 nm. All nitride film depositions were carried out with 20 sccm Ar/5sccm N$_2$ gas flow and a target-to-substrate distance of 60 mm. The typical film thickness was 2.5–3 µm.

Film compositions were analyzed using X-ray photoelectron spectroscopy (XPS) on a Kratos Axis/HS system (XPS Axis HSi, Kratos Analytical, Manchester, UK). Samples were Ar$^+$-ion etched before analysis to remove surface contaminants. The atomic percentages of nitrogen and oxygen were determined along with the metallic elements in 304 stainless steel (Fe, Ni, and Cr). The accuracy in nitrogen concentration measurements is estimated to be ±(2–4) at.% N. X-ray diffraction analysis of the films was first carried out using a Shimadzu XRD-6100 (Shimadzu, Columbia, MD, USA) using CuKα radiation (λ = 0.1542 nm) set up in the Bragg-Brentano configuration with a graphite diffracted-beam monochromator. Additional X-ray diffraction studies were carried out using a Bruker system (Bruker Inc., Madison, WI, USA) equipped with a Vantec-500 area detector. The goniometer used was equipped with a two-position χ stage, which for the present purposes was set at the χ = 54.74° position. The nominal detector distance was set at 8 cm, and a CoKα X-ray tube was used in order to avoid fluorescence of Fe. The accuracy in the measurement of interplanar spacings for this system is discussed in Appendix A. Further details on the analysis of the area detector data are outlined in the following section.

Samples were deposited with substrate temperature and bias as the experimental variables. The temperatures used were between 150 and 350 °C, while the substrate bias was set at either ground or a level between −40 and −160V. (Several samples were deposited at room temperature but had an amorphous structure and are therefore not considered here.) In the discussion which follows, the samples are denoted by temperature and bias (with zero bias indicating ground), e.g., sample S-150-60 indicates deposition using the 304 stainless steel target and 150 °C and −60 V bias.

3. Results

3.1. Film Compositions

The relative proportions of Fe, Ni and Cr found in the films generally reflected those of the target material, which for 304 stainless steel is nominally 74% Fe, 18% Cr, and 8% Ni. The films also contained some oxygen, for samples deposited with a bias the average was 3.9 at.%, while for samples deposited at ground the average was 16.2%. The nitrogen levels in the films are shown in Figure 2 as a function of substrate temperature and bias. Two general trends are observed: first, at a given temperature, higher substrate bias levels results in a lower nitrogen content; this could be due to sputtering of nitrogen during deposition. This concept is supported by the fact that samples deposited at ground had the highest nitrogen content. Substrate temperature had less of an effect, mostly resulting in a in a slightly higher nitrogen level at higher temperatures. However, these trends were not significantly larger than the accuracy of the measurement.

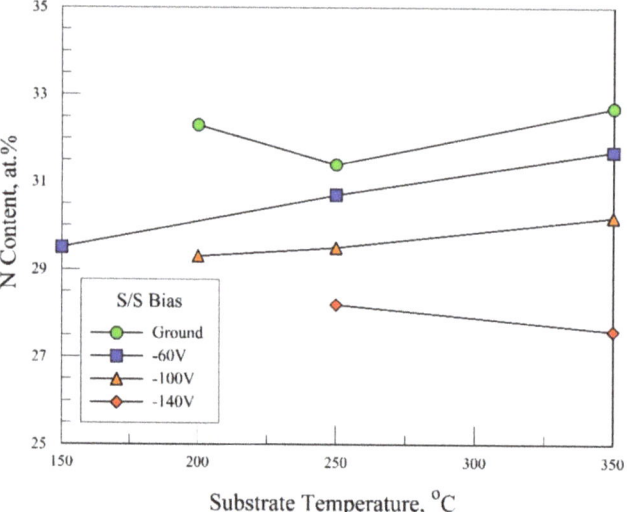

Figure 2. Nitrogen concentration in films deposited at different substrate temperature and bias levels. The major effect on composition is due to changes in substrate bias.

3.2. X-ray Diffraction Using the Bragg-Bretano Method

Figure 3a–d shows X-ray diffraction results collected on the Shimadzu diffractometer, which operates in the Bragg-Bretano configuration. Spectra are shown for the range of $2\theta = 30°$ to $65°$; the latter limit was chosen to avoid the highly intense Si (400) substrate peak which appears at $2\theta = 69.20°$. This also obstructed detection of the (220) peaks in the films, however, as verified later using the area detector XRD system, these peaks were either very weak or absent due to the film texture. Additional spectra were recorded in the range of $2\theta = 70°$ to $120°$, but again peaks in this range (primarily the (311) and (222) reflections) were generally weak and not used in the analysis.

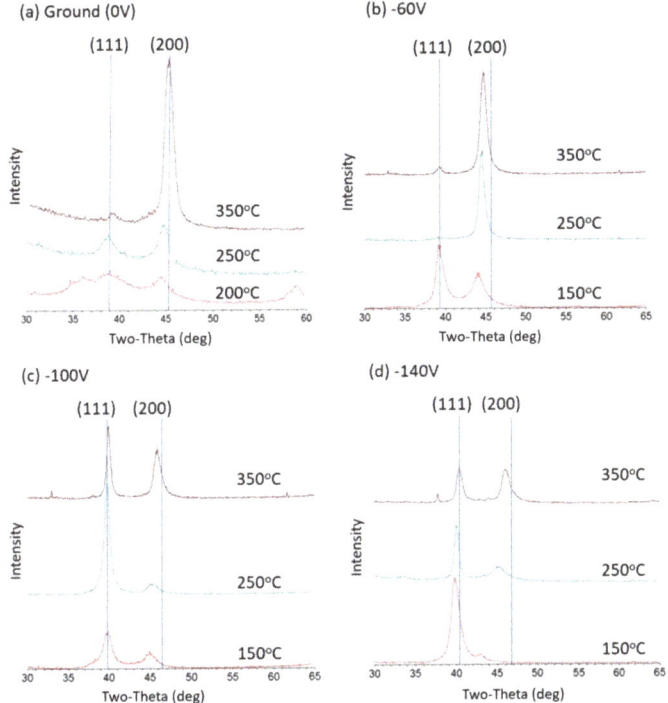

Figure 3. X-ray diffraction results for deposited films collected using the Bragg-Bretano configuration and CuKα radiation. The spectra are shown for substrate bias levels of (**a**) 0 V, (**b**) −60 V, (**c**) −100 V and (**d**) −140 V, each at three different temperatures as shown. The positions of the (111) and (200) lines are shown and their locations based on a method described in the text.

Since there is no X-ray diffraction standard for S-phase structured stainless steel nitrides, analysis of the experimental patterns shown in Figure 3 require that an initial assumption be made to determine the appropriate lattice constant. Typically, the (200)/(400) peaks are assumed to have the anomalous shift so the (111) is used to determine the lattice constant (denoted a_{111}) and then remaining peak positions are determined based on this value and the assumption of an ideal FCC structure. This analysis was carried out by averaging the a_{111} values for each set of films at the given bias and then displaying the (111) and (200) positions for each group in Figure 3.

The results show that the deviation in the (200) peak, compared to its expected position based on an ideal FCC structure, is typically ~1°–1.5°–2θ lower. For samples deposited at ground (0 V) the peaks are broad, typically an effect of poor crystallinity; the sample deposited at 150 °C was too poorly crystalline for useful analysis and so a 200 °C deposition was conducted instead. In comparison, the −60 V samples show better crystallinity and a larger (200) peak shift. Figure 3d (−140 V bias) also shows for the 150 °C sample a small peak at 42.86°. While this could be a highly shifted (200) reflection, a more plausible explanation is that it is the result of small amount of Cr_2N formation, which according to PDF#35-0803 has a (111) peak at 42.61°. The formation of this phase, not seen in other samples, is possibly due to the high bias and low temperature used as deposition conditions for this sample.

The XRD data in Figure 3 was analyzed and measurements of θ_{111} and θ_{200} were obtained allowing calculation of the R_B parameter described in Equation (1). The results are shown in Figure 4. It can be noted that the R-values are all greater than 0.75. The effect of increasing substrate temperature at a given bias is to generally reduce the value of R_B, indicating a more normal cubic structure. The effect of substrate bias shows an increase of R_B with bias level. The effect of bias was examined by closely by depositing samples at a constant temperature of 250 °C and varying the bias levels from 0 to −160V. The results are shown in Figure 5, where the data has also been fitted to a parabolic curve. For the sample deposited at 0V, the structure is very close to the ideal cubic structure, but at −160 V a very large value of R_B = 0.795 is obtained. Therefore, based on the results shown in Figures 3 and 4, it can be concluded that lower temperatures and higher bias levels promote a larger deviation from the peak positions expected from a standard cubic structure.

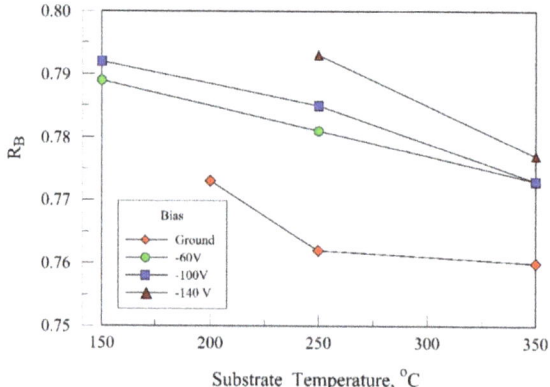

Figure 4. Measured values of R_B for films deposited at various bias levels and substrate temperatures. The values of R_B generally decrease with temperature and increase with substrate bias.

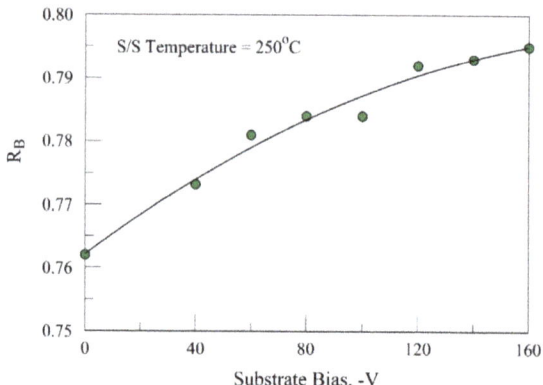

Figure 5. R_B values for films deposited at 250 °C and substrate bias levels ranging from 0 to −160 V. The line shown is a parabolic fit to the data.

3.3. Area-Detector Diffraction Studies

Additional X-ray diffraction studies were carried out using a Bruker system equipped with a Vantec-500 area detector and a CoKα radiation source. By setting the χ-stage in the 54.74° position, we were able to collect X-ray data in the range of ψ ~0° to 55° (with ψ = 0° being the normal Bragg-Brentano position). However, in some of the films, the presence of significant texture did not allow accurate measurements of peak positions at all ψ angles. Therefore, only a subset of the deposited films could be analyzed by this technique, as indicated below.

In order to analyze samples, a set of four raw frames (at 90° rotations about the sample normal) was first collected on the Vantec-500 detector which was set at a 2θ angle to allow optimal position of the two partial Debye rings corresponding to the (111) and (200) lines. The 2θ-peak positions (vs. ψ) for each ring were determined by using the Bruker GADDS stress analysis software (v. 4.1.51) for bi-axial 2D analysis, which divides the ring into 10 segments and integrates each segment (0.1° step size) to determine the average peak position for that segment. The peak positions were converted to d-spacings, giving the d_{200} and d_{111} values vs. ψ angle. The next step in the analysis was to obtain tables of d_{111} and d_{002} values as a function of φ, which is the angle between the (002) direction (c-axis) and the substrate plane. For d_{002} values the ψ and φ angles are the same, but for (111) reflections, the conversion φ = 54.74 − ψ was necessary, as shown in Figure 1. The d_{111} values were then reordered to list the results in increasing values of φ, from near zero to approximately 55°. While the angular range for φ was similar for the d_{111} and d_{200} data sets, the average angle of each of the 10 integrated sections was not the same, so for further analysis the data were fit to a second-order polynomial ranging from φ = 0° to 60° in 5° steps. The fitted d_{111} and d_{002} values were first used to calculate the a_{111} and a_{002} values based on the assumption of a cubic lattice. They were then used to calculate the parameters for a tetragonal lattice, a_{200} (=a) and a_{002} (=c), using the relation $c = 2 d_{002}$ and the a value obtained from Equation (3). In addition, the volume of the tetragonal unit cell was calculated using the equation $V = a^2 c$.

Figure 6 plots the values of a_{111} and a_{200} based on a cubic lattice assumption. As expected, the a_{200} values are always higher than a_{111}. The degree of separation between the curves is consistent with Figure 4. The variations of a_{hkl} with φ provides important information on possible residual stress effects and will be discussed further in the following section. Comparing actual R_B values from the area detector data (at ψ = 0) with those in Figure 4, it was found that the area detector results gave R_B values 5%–6% higher in all four cases. Figure 6 also shows that, in all cases, the a-values decrease with increasing φ, although not at the same rate for all samples.

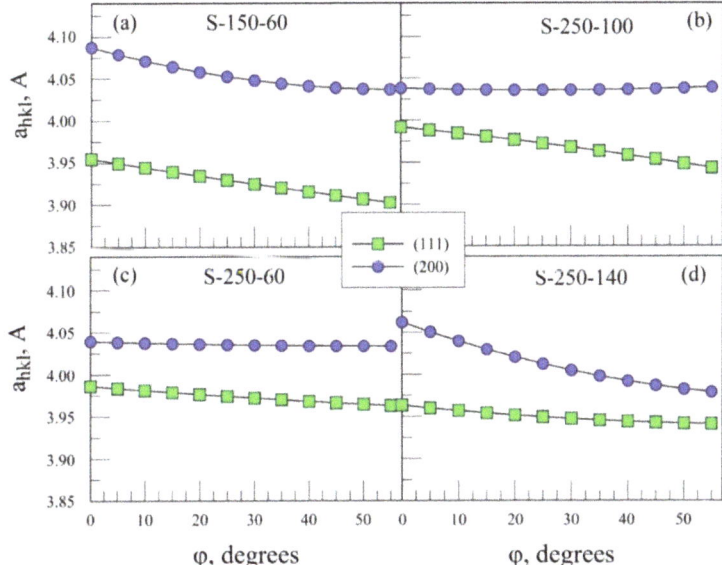

Figure 6. Lattice constant values as calculated from the positions of the (111) and (200) peaks, shown for four different deposition conditions (**a**–**d**). The lattice constants shown here were calculated based on the assumption of a cubic structure. The values are shown as a function of φ; at φ = 0 the [002] direction is parallel to the surface plane normal.

Next, assuming a tetragonal lattice, the c/a values were calculated and the results are shown in Figure 7. Overall, the values range from about 1.01 to 1.05. The φ-dependence is varied, ranging from almost constant (S-150-60) to strongly decreasing (S-250-140) to increasing (S-250-100). The values tend to show as slight decrease in *c/a* near 30°, but this has be determined to be due to a slight (and non-correctable) misalignment of the detector. Figure 8 shows the unit cell volume as a percentage change from the initial (φ = 0) volume. In all cases, the unit cell volume decreases with increasing φ, with the maximum values shown (at φ = 55°) ranging from about −1% to −4%. Sample S-150-60 showed the largest percentage decrease, and was also the sample with the highest c/a values.

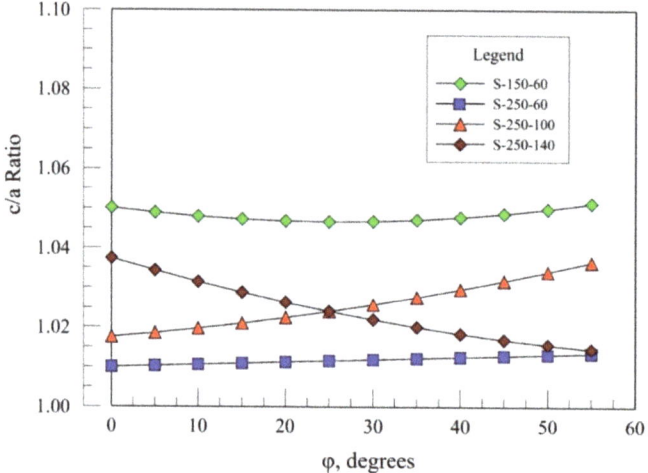

Figure 7. Calculated *c/a* ratio as determined from area-detector diffraction analysis. The *c/a* ratios are all greater than 1.0 but the dependence on φ varies considerably in both magnitude and rate of change with φ.

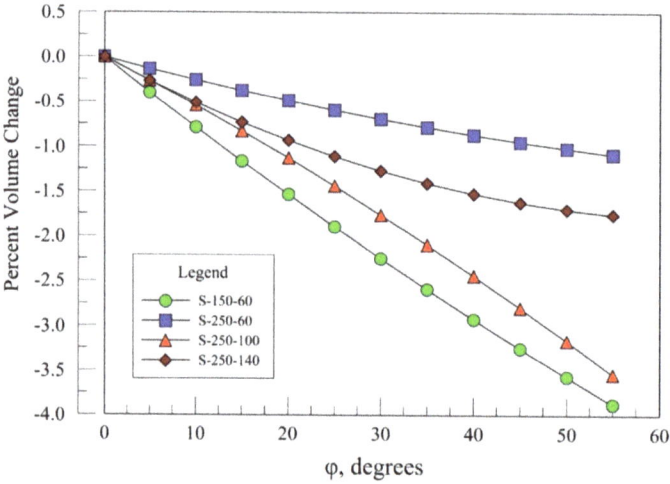

Figure 8. Percentage volume change for the tetragonal lattice based on area detector X-ray diffraction data. In all cases, the volume decreased as the φ angle increased. The decrease is most significant for sample S-150-60, which had the highest *c/a* ratio.

Having calculated c/a ratios for these samples, we can calculate the expected separation between $2\theta_{002}$ and $2\theta_{002}$, denoted here as $\Delta 2\theta$. Values were calculated for the four samples examined in this section, and a range of values was obtained depending of the ψ values. The results are shown in Table 1. In general, even for broad peaks of the nature shown in Figure 4, the (002)/(200) peak split is large enough in most cases to be readily observed. Yet, examination of the entire area detector field showed no extra peaks were present.

Table 1. Expected for (200)/(002) peak split.

Sample	Δ2θ Range
S-60-150	2.8–2.9°
S-60-250	0.6–0.8°
S-100-250	1.4–2.2°
S-140-250	0.84–2.1°

4. Discussion

In this study the nature of the well-known diffraction anomaly observed in samples of expanded austenite or the S-phase in nitrogen supersaturated stainless steels has been studied in sputter-deposited thin films. The use of this thin film deposition technique allows us to examine the effects of variations in the substrate temperature and bias, and films can be deposited at temperatures lower than those typically used in plasma nitriding of bulk samples. Also, in contrast to most plasma-nitriding methods, sputter deposition results in compositionally uniform samples, and therefore simplifies the analysis of X-ray diffraction data.

The extent of the diffraction anomaly was first characterized by calculating "R_B-values" from standard Bragg-Bretano X-ray diffraction patterns, with $R_B = 0.75$ indicating a normal cubic lattice. As shown in Figures 4 and 5, the R_B-values increased with substrate bias and decreased with increasing temperature. These results can be considered in parallel with the nitrogen content in the films (Figure 2), which decreases with substrate bias but where only a small temperature effect is observed. The increase in R_B with bias, as well as the decrease in nitrogen content indicates that less N in the lattice increases the diffraction anomaly. Kappaganthu and Sun [15] deposited similar films by reactive sputtering using a range of nitrogen gas concentrations and obtained R_B-values (as calculated based on their data) similar to those reported here, but also obtained stoichiometric films which had R_B-values of 0.75. These results support the conclusion that the extent of the diffraction anomaly is proportional to the deficiency of nitrogen in the lattice. However, the results shown in Figure 4 also show a trend of decreasing R_B with increasing temperature in samples that had a relatively constant N content. It is well established, via the zone model, that higher deposition temperatures increases the film density and grain size while reducing film defects such as intergranular voids, faceted columns, and feather-like structures [37]. However, the presence of such defects is not known to produce a diffraction anomaly. In addition, the results shown in Figure 4 appear to suggest contradictory evidence for this hypothesis: R_B decreases with increasing substrate deposition temperature, which should help densify films, but increases with bias, which also increases film density. The possibility of peak shifts due to stacking fault defects, as discussed in the Introduction, may explain the temperature effect but detailed microscopic examinations of our samples will be needed to verify this.

One explanation for the diffraction anomaly is that (002) and (111)-oriented grains have different nitrogen concentrations, and therefore different lattice constants; when measured using the Bragg-Brentano method each peak would represent different grains. Therefore, it is important to obtain the (111) and (002) interplanar spacings form similarly-oriented grains. This was done here using the area detector diffraction method, where interplanar spacings from planes oriented away from the substrate surface orientation could be measured. This allowed the interplanar spacings of (111) and (002) planes to be measured for grains of similar orientations. The results were first analyzed assuming a cubic structure, as shown in Figure 6. As expected, the a_{111} and a_{200} had different values over the range of φ values. Generally, these a_{hkl} either decreased or remained constant with φ but were still significantly different over the range of φ. Physically, this means that measured a_{111} and a_{200} values were truly different within a single grain and not due to the fact that each was measure from grains of different orientations.

The variations of a_{hkl} with φ observed in Figure 6 can be considered as possible effects of residual stress. We first consider the (002) planes parallel to the substrate normal, as shown in grain 1 in Figure 1. For a compressive (in-plane) residual stress, these planes would have a higher value of a_{002} compared to the unstressed state. For (002) planes tilted away from the substrate normal (increasing φ), as shown in grain 2, the value of a_{002} will decrease. This is observed for Figure 6a,d (the curves in Figure 6b,c suggest little or no stress is present in these cases). For the (111) planes, the φ = 0 case refers to the tilted case (as shown in grain 1) and as φ increases the (111) planes become increasingly parallel to the surface. Therefore, when plotting the data as a vs. φ, the a_{111} values should increase with increasing φ (for a compressive stress). This is contrary to the data shown in Figure 6, where a_{111} is always decreasing. Therefore, while residual stress may impact the curves, it alone cannot explain the data shown in Figure 6.

Next, the possibility of explaining the peak shift using a tetragonal structure was explored. Using Equation (3), the c/a values were calculated as shown in Figure 7. The values were all greater than one, however, no consistent trend was observed when examining the φ-dependence. In fact, results ranged from being relatively constant to increasing to decreasing. However, calculation of relative unit cell volume (Figure 8) did show a consistently decreasing value with φ although with varying magnitudes. This is consistent with the observation that the diffusion of N is highest for (002) oriented grains, giving these grains a higher N concentration and larger lattice constant. Even without the tetragonal lattice assumption (as observed in the data in Figure 6), the largest lattice constants are generally found in the [002]-oriented grains.

The tetragonal model can also be used to calculate the location of the additional peaks that should be observed, and Table 1 shows calculations of the expected peak split for the (002)/(200) reflections. However, no additional peaks were observed. Despite the fact that the d-spacings from the (111) and (002) peaks are inconsistent with the cubic structure, no evidence for a non-cubic structure could be found. This suggests that the anomaly is related to defects created by sub-stoichiometric N content, which is supported by the decreasing R_B values with increasing substrate temperature.

5. Conclusions

Thin films of nitrogen-enhanced 304 stainless steels were deposited by magnetron sputtering using primarily substrate temperature and bias as deposition variables. Samples were analyzed using X-ray diffraction methods in order to help understand the origin of the well-known diffraction anomaly commonly observed in S-phase samples.

The films were found to contain between 28 and 32 at.% nitrogen. Higher substrate bias levels results in a lower nitrogen contents, possibly due to sputtering of nitrogen during deposition. Substrate temperature had only a minor effect, mostly resulting in a in a slightly higher nitrogen level at higher temperatures.

X-ray diffraction using the Bragg-Brentano method was carried out with particular focus on measuring the positions of the (111) and (002) peaks. Using these results a term denoted R_B, which is related to the extent of the diffraction anomaly, was calculated as a function of deposition conditions. It was found that the R_B values decreased with substrate temperature, and increased strongly with substrate bias.

Area detector diffractometry studies were conducted and analyzed using a tetragonal structure model. This allowed calculation of c/a ratios and the expected (002)/(200) peak split. However, no peak split was observed, indicating the tetragonal structure model is not valid for these samples. This also suggests that a defect-based hypothesis is more viable as an explanation for the diffraction anomaly.

Author Contributions: Formal analysis, F.I.A. and J.E.K.; Investigation, F.I.A.; Methodology, J.E.K.; Resources, J.E.K.; Writing—original draft, F.I.A.; Writing—review & editing, J.E.K. All authors have read and agreed to the published version of the manuscript.

Funding: This research received no external funding.

Conflicts of Interest: The authors declare no conflict of interest.

Appendix A

The discrepancy between the a_{111} and a_{200} measured lattice constants, as shown for example in Figure 6, needs to be considered in comparison to the accuracy typical of XRD measurements. To examine this more closely, we measured the a_{111}/a_{200} lattice constants for a Cu powder, for which the calculated lattice constants should be identical. Data were acquired in a manner similar to that for Figure 6 (Using the Bruker XRD instrument) and processed in a similar way to obtain a vs. φ. The results are shown below in Figure A1, where the scale for the a-values was chosen to be similar to that in Figure 6. The discrepancy is at most ~0.02 A, whereas the difference between a_{111} and a_{200} in Figure 6 is typically between 0.05–0.1 A. This supports the fact that the a_{111}/a_{200} lattice constant differences shown in Figure 6 is not due to measurement inaccuracies.

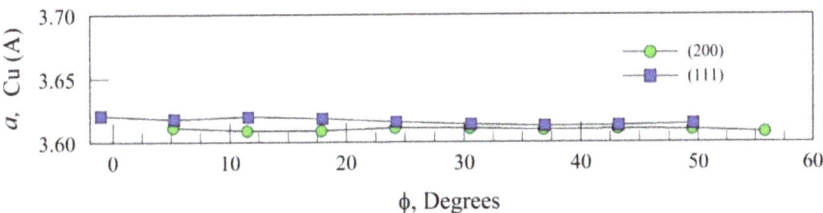

Figure A1. Measured lattice constants for a Cu powder based on the (111) and (200) reflections vs. φ, angle increased. The figure demonstrates the typical accuracy of the lattice constant measurement.

References

1. Zhang, Z.L.; Bell, T. Structure and corrosion resistance of plasma nitrided stainless steels. *Surf. Eng.* **1985**, *1*, 131–136. [CrossRef]
2. Ichii, K.; Fujimura, K.; Takase, T. Sturcture of the ion-nitrided layer of 18-8 stainless steel. *Technol. Rep. Kansai Univ.* **1986**, *27*, 135–144.
3. Ozturk, O.; Williamson, D.L. Phase and composition depth distribution analyses of low energy, high flux N implanted stainless steel. *J. Appl. Phys.* **1995**, *77*, 3839–3850. [CrossRef]
4. Williamson, D.L.; Davis, J.A.; Wilber, P.J.; Vajo, J.J.; Wei, R.; Matossian, J.N. Relative roles of ion energy, ion flux, sample temperature in low-energy nitrogen ion implantation of Fe-Cr-Ni stainless steel. *Nucl. Inst. Methods Phys. Res. B* **1997**, *127–128*, 930–934. [CrossRef]
5. Wei, R.; Vajo, J.J.; Matossian, J.N.; Wilbur, P.J.; Davis, J.A.; Williamson, D.L.; Collins, G.A. A comparative study of beam ion implantation, plasma ion implantation, and nitriding of AISI 304 stainless steel. *Surf. Coat. Technol.* **1996**, *83*, 235–242. [CrossRef]
6. Samandi, M.; Shedden, B.A.; Smith, D.I.; Collins, G.A.; Hutchings, R.; Tendys, J. Microstructure, corrosion, and tribological behaviour of plasma immersion ion-implanted austenitic stainless steel. *Surf. Coat. Technol.* **1993**, *59*, 261–266. [CrossRef]
7. Menthe, E.; Rie, K.-T.; Schultze, J.W.; Simson, S. Structure and properties of plasma-nitrided stainless steel. *Surf. Coat. Technol.* **1995**, *74–75*, 412–416. [CrossRef]
8. Mändl, S.; Günzel, R.; Richter, E.; Möller, W. Nitriding of austenitic stainless steels using plasma immersion ion implantation. *Surf. Coat. Technol.* **1998**, *372*, 100–101. [CrossRef]
9. Blawert, C.; Mordike, B.L. Nitrogen plasma immersion ion implantation for surface treatment and wear protection of austenitic stainless steel X6CrNiTi1810. *Surf. Coat. Technol.* **1999**, *116–119*, 352–360. [CrossRef]

10. Leyland, A.; Lewis, D.B.; Stevenson, P.R.; Matthews, A. Low temperature plasma diffusion treatment of stainless steels for improved wear resistance. *Surf. Coat. Technol.* **1993**, *62*, 608–617. [CrossRef]
11. Saker, A.; Leroy, C.; Michel, H.; Frantz, C. Properties of sputtered stainless-steel coatings and structural analogy with low temperature plasma nitride layers of austenitic steels. *Mater. Sci. Eng.* **1991**, *A140*, 702. [CrossRef]
12. Bourjot, A.; Foos, M.; Frantz, C. Basic Properties of Sputtered 310 Stainless Steel-Nitrogen Coatings. *Surf. Coat. Technol.* **1990**, *43–44*, 533–542. [CrossRef]
13. Shedden, B.A.; Kaul, F.N.; Samandi, M.; Window, B. The role of energetic neutrals in reactive magnetron sputtering of nitrogen-doped austenitic stainless steel coatings. *Surf. Coat. Technol.* **1997**, *97*, 102–108. [CrossRef]
14. Terwagne, G.; Hody, H.; Colaux, J. Structural and quantitative analysis of stainless steel coatings deposited by DC-magnetron sputtering in a reactive atmosphere. *Surf. Coat. Technol.* **2002**, *383*, 174–175. [CrossRef]
15. Kappaganthu, S.R.; Sun, Y. Formation of an MN-type cubic nitride phase in reactively sputtered stainless steel-nitrogen films. *J. Cryst. Growth* **2004**, *267*, 385–393. [CrossRef]
16. Kappaganthu, S.R.; Sun, Y. Influence of sputter deposition conditions on phase evolution in nitrogen doped stainless steel films. *Surf. Coat. Technol.* **2004**, *198*, 59–63. [CrossRef]
17. Marchev, K.; Hidalgo, R.; Landis, M.; Vallerio, R.; Cooper, C.V.; Giessen, B.C. The metastable *m* phase layer on ion-nitrided austenitic stainless steels Part 2: Crystal structure ad observation of its two-directional orientation anisotropy. *Surf. Coat. Technol.* **1999**, *67*, 112. [CrossRef]
18. Marchev, K.; Landis, M.; Vallerio, R.; Cooper, C.V.; Giessen, B.C. The *m* phase layer on ion-nitrided austenitic stainless steels (III): An epitaxial relationship between the *m* Phase and the γ parent phase and a review of structural identification of this phase. *Surf. Coat. Technol.* **1999**, *116*, 184–188. [CrossRef]
19. Bacci, T.; Borgioli, F.; Galvanetto, E.; Pradelli, G. Glow-discharge nitriding of sintered stainless steels. *Surf. Coat. Technol.* **2001**, *139*, 251–256. [CrossRef]
20. Fewell, M.P.; Mitchell, D.R.G.; Priest, J.M.; Short, K.T.; Collins, G.A. The nature of expanded austenite. *Surf. Coat. Technol.* **2000**, *131*, 300–306. [CrossRef]
21. Fewell, M.P.; Priest, J.M. Higher-order diffractometry of expanded austenite using synchrotron radiation. *Surf. Coat. Technol.* **2008**, *202*, 1802. [CrossRef]
22. Blawert, C.; Kalvelage, H.; Mordike, B.L.; Collings, G.A.; Short, K.T.; Jirásková, Y.; Schneeweiss, O. Nitrogen and carbon expanded austenite produced by PI^3. *Surf. Coat. Technol.* **2001**, *136*, 181–187. [CrossRef]
23. Christiansen, T.; Somers, M.A.J. On the crystallographic structure of the S-phase. *Scr. Mater.* **2004**, *50*, 35–37. [CrossRef]
24. Xu, X.L.; Yu, Z.W.; Wang, L.; Hei, Z.K. Microstructural characterization of plasma nitride austenitic stainless steel. *Surf. Coat. Technol.* **2000**, *132*, 270–274. [CrossRef]
25. Warren, B.E. *X-ray Diffraction*; Addison-Wesley: Boston, MA, USA, 1969; pp. 275–298.
26. Xu, X.L.; Yu, Z.W.; Wang, L.; Qiang, J.; Hei, Z.K. Phase depth distribution characteristics of the plasma nitrided layer on AISI 304 stainless steel. *Surf. Coat. Technol.* **2003**, *162*, 242–247. [CrossRef]
27. Stroz, D.; Psoda, M. TEM studies of plasma nitride austenitic stainless steel. *J. Microsc.* **2010**, *237*, 227–231. [CrossRef]
28. Brink, B.K.; Stahl, K.; Christiansen, T.L.; Oddershede, J.; Winther, G.; Somers, M.A.J. On the elusive structure of expanded austenite. *Scr. Mater.* **2017**, *131*, 59–62. [CrossRef]
29. Velterop, L.; Delhez, R.; deKeijser, T.H.; Mittemeijer, E.J.; Reefman, D. X-ray diffraction analysis of stacking and twin faults in f.c.c. metals: A revision and allowance for texture and non-uniform fault probabilities. *J. Appl. Cryst.* **2000**, *33*, 296–306. [CrossRef]
30. Grigull, S.; Parascandola, S. Ion-intriding induced plastic deformation in austenitic stainless steel. *J. Appl. Phys.* **2000**, *88*, 6925–6927. [CrossRef]
31. Abrasonis, G.; Riviere, J.P.; Templier, C.; Declemy, A.; Pranevicius, L.; Milhet, X. Ion beam nitriding of single and polycrystalline austenitic stainless steel. *J. Appl. Phys.* **2005**, *97*, 083531. [CrossRef]
32. Brink, B.K.; Stahl, K.; Christiansen, T.L.; Fradsen, C.; Hansen, M.F.; Somers, M.A.J. Composition-dependent variation of magnetic properties and interstitial ordering in homogeneous expanded austenite. *Acta Mater.* **2016**, *106*, 32–39. [CrossRef]
33. Oddershede, J.; Christiansen, T.L.; Stahl, K.; Somers, M.A.J. EXAFS investigation of low temperature nitride stainless steel. *J. Mater. Sci.* **2008**, *43*, 5358–5367. [CrossRef]

34. Czerwiec, T.; Andrieux, A.; Marcos, G.; Michel, H.; Bauer, P. Is "expanded austenite" really a solid solution? Mössbauer observation of an annealed AISI 316L nitride sample. *J. Alloys Compd.* **2019**, *811*, 151972. [CrossRef]
35. Christiansen, T.L.; Hummelshøj, T.S.; Somers, M.A.J. Expanded austenite, crystallography and residual stress. *Surf. Eng.* **2010**, *26*, 242–247. [CrossRef]
36. Alresheedi, F.I.; Krzanowski, J.E. Structure and morphology of stainless steel coatings sputter-deposited in a nitrogen/argon atmosphere. *Surf. Coat. Technol.* **2017**, *314*, 105–112. [CrossRef]
37. Mayrhofer, P.H.; Mitterer, C.; Hultman, L.; Clemens. H. Microstructural design of hard coatings. *Prog. Mater. Sci.* **2006**, *51*, 1032–1114. [CrossRef]

Publisher's Note: MDPI stays neutral with regard to jurisdictional claims in published maps and institutional affiliations.

© 2020 by the authors. Licensee MDPI, Basel, Switzerland. This article is an open access article distributed under the terms and conditions of the Creative Commons Attribution (CC BY) license (http://creativecommons.org/licenses/by/4.0/).

Article

Examination of the Hydrogen Incorporation into Radio Frequency-Sputtered Hydrogenated SiN$_x$ Thin Films

Nikolett Hegedüs [1,2,3], Riku Lovics [1], Miklós Serényi [1], Zsolt Zolnai [1], Péter Petrik [1], Judit Mihály [4], Zsolt Fogarassy [1], Csaba Balázsi [1] and Katalin Balázsi [1,*]

[1] Centre for Energy Research, Institute for Technical Physics and Materials Science, Konkoly-Thege M. Str. 29-33, 1121 Budapest, Hungary; nhegedus@guardian.com (N.H.); drlovicsriku@gmail.com (R.L.); serenyi.miklos@energia.mta.hu (M.S.); zolnai.zsolt@ek-cer.hu (Z.Z.); petrik.peter@ek-cer.hu (P.P.); fogarassy.zsolt@ek-cer.hu (Z.F.); balazsi.csaba@ek-cer.hu (C.B.)
[2] Doctoral School of Materials Science and Technologies, Óbuda University, Bécsi Str. 96/B, 1030 Budapest, Hungary
[3] Guardian Orosháza Ltd., Csorvási u. 31, 5900 Orosháza, Hungary
[4] Research Centre for Natural Sciences, Institute of Materials and Environmental Chemistry, Magyar Tudósok Krt. 2, 1117 Budapest, Hungary; mihaly.judit@ttk.mta.hu
* Correspondence: balazsi.katalin@ek-cer.hu

Abstract: In this work, amorphous hydrogen-free silicon nitride (a-SiN$_x$) and amorphous hydrogenated silicon nitride (a-SiN$_x$:H) films were deposited by radio frequency (RF) sputtering applying various amounts of hydrogen gas. Structural and optical properties were investigated as a function of hydrogen concentration. The refractive index of 1.96 was characteristic for hydrogen-free SiN$_x$ thin film and with increasing H$_2$ flow it decreased to 1.89. The hydrogenation during the sputtering process affected the porosity of the thin film compared with hydrogen-free SiN$_x$. A higher porosity is consistent with a lower refractive index. Fourier-transform infrared spectroscopy (FTIR) confirmed the presence of 4 at.% of bounded hydrogen, while elastic recoil detection analysis (ERDA) confirmed that 6 at.% hydrogen was incorporated during the growing mechanism. The molecular form of hydrogen was released at a temperature of ~65 °C from the film after annealing, while the blisters with 100 nm diameter were created on the thin film surface. The low activation energy deduced from the Arrhenius method indicated the diffusion of hydrogen molecules.

Keywords: SiN$_x$:H; refractive index; activation energy; structure

1. Introduction

Hydrogenated silicon nitride (SiN$_x$:H) films are widely used in the microelectronics industry to enhance the efficiency of silicon-based light emitters [1] or to improve the efficiency of silicon solar cells as the antireflective and passivation layer on the front surface of such device structures. Silicon nitride (Si$_3$N$_4$, hereinafter referred to as SiN) thin films may be applied as inorganic gate insulators in organic thin film transistors (OTFTs) [2]. SiN:H is also an appropriate material for charge trap functional region of non-volatile memory (NVM) structures [3]. Their tunable refractive index together with the low extinction coefficient enable the application of SiN$_x$ as an excellent antireflection thin film [4]. Furthermore, SiN$_x$:H containing multilayer stacks with a gradient refractive index profile were recommended to further decrease the optical losses [5]. Apart from the optical properties, the passivation effect of SiN$_x$:H layers is also substantial since recombination losses significantly restrict the performance of solar cells. Hydrogenated silicon nitride films have been reported to show a good surface and bulk passivation effect after annealing due to atomic hydrogen diffusion to the surface [6]. As dangling bonds are healed by hydrogen, the number of recombination centers can be reduced, which results in increasing carrier lifetime. However, the molecular hydrogen migration requires a higher activation energy and low diffusivity [7].

The most common techniques for deposition of the silicon nitride films with or without hydrogen addition are different types of chemical vapor deposition (CVD), such as plasma-enhanced chemical vapor deposition (PECVD) [8,9], remote plasma-enhanced CVD (RPECVD) [10], electron cyclotron resonance (ECR) [11,12], hot-wire CVD (HWCVD) [13], and extended thermal plasma CVD (ETPCVD) [14]. CVD-deposited film always contains hydrogen but its amount cannot be directly controlled during the preparation process, only by several deposition parameters, such as the ratio of precursor gases or the substrate temperature [9,12]. Due to this fact, the magnetron sputtering technique could be the alternative fabrication method for directly controlled hydrogen concentration via adjusting the applied hydrogen gas flow to the chamber [15].

Direct current (DC) magnetron sputtering [16], radio frequency (RF) sputtering [17], and high-power impulse magnetron sputtering (HiPIMS) [18] were also proved to be proper methods to produce SiN_x:H thin films at a lower substrate temperature. In the case of different sputtering techniques, it is possible to directly control the amount of hydrogen by adjusting the applied hydrogen gas flow. K. Mokeddem et al. developed the DC magnetron sputtering (DCMS) technique to prepare hydrogenated amorphous silicon nitride thin films in argon gas flow mixed with molecular hydrogen and nitrogen [16]. These films presented a large band gap and showed a nearly stoichiometric composition, and both nitrogen and hydrogen were incorporated into the structure. F.L. Martinez et al. used electron cyclotron resonance plasma-enhanced CVD for amorphous hydrogenated silicon nitride (a-SiN_x:H) film deposition under different values of gas flow ratio, deposition temperature, and microwave power [12]. The formation of Si–H and N–H as competitive processes occurred during the film growth. They showed that the substitution of N–H bonds with Si–H bonds was driven by the tendency for chemical order or maximum bonding energy. V. Tiron et al. used the reactive HiPIMS technique to fabricate SiN_x:H thin films. Their coating showed a very smooth surface with a dense homogeneous amorphous and amorphous to nanocrystalline structure [18]. This coating revealed a diffusion process of atomic H into the Si substrate, indicating the presence of numerous hydrogen bonds (Si–H and N–H) that could passivate structural defects and reduce the number of recombination centers in silicon bulk.

The physical, electrical, and optical properties of amorphous SiN_x and SiN_x:H thin films strongly depend on film composition and the applied deposition technique. Ellipsometry is a widely used non-destructive technique for the optical characterization of a wide range of thin layer-on-substrate material systems. The optical properties cannot be deduced directly from the raw measurement data but indirectly from the ellipsometric modeling process. Therefore, the choice of the applied ellipsometric model is an important point of data evaluation. Boulesbaa et al. developed an effective medium approximation (EMA) model, which considers SiN_x:H films as a mixture of silicon (Si), stoichiometric silicon nitride, and hydrogen (H_2) [19]. In addition to experimental work, the optical properties of SiN:H films have also been studied by theoretical models. F. de Brito Mota et al. developed an interatomic potential to describe a-SiN_x:H thin films of varying nitrogen content [20]. They found that hydrogen incorporation into silicon nitride films leads to the reduction of dangling bonds corresponding to undercoordinated silicon and nitrogen atoms. Tao et al. calculated the optical properties of SiN_x:H films based on the density functional theory [21]. They found that the hydrogen incorporation into the silicon nitride films had a healing effect by saturating the dangling bonds, which leads to the decrease of absorption coefficient and refractive index of the films.

Different deposition conditions may result in modified coating structure and material properties. In this work, amorphous hydrogen-free (a-SiN_x) and a-SiN_x:H thin films were deposited by RF sputtering onto two kinds of substrates (Si (001) and glass) with various amounts (0–12 sccm) of hydrogen. The effects of the hydrogen flow on the optical and structural properties were investigated.

2. Experimental Section

Both types of silicon nitride thin films (a-SiN$_x$, a-SiN$_x$:H) were deposited by a Leybold Z400 Radio Frequency Sputtering (RFS) tool (Figure 1). The base pressure was 2×10^{-5} mbar. A circular Si target with a diameter of ~76 mm (Kurt J. Lesker Comp., undoped with 99.99% purity) and N$_2$ gas source were applied for deposition of the hydrogen-free silicon nitride films.

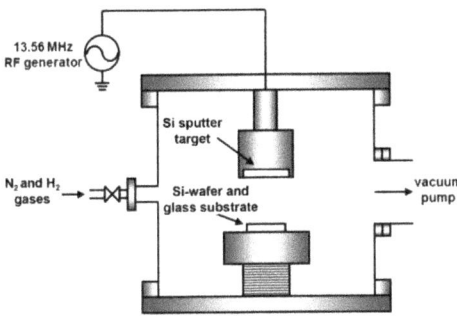

Figure 1. Schematic view of the sputtering system.

In the case of SiN$_x$:H films, N$_2$ and hydrogen with a flow rate in the range from 0 to 12 sccm were added. The gas flow rates were controlled by adjustable mass flow controllers (MFCs). The sputtering parameters are summarized in Table 1.

Table 1. Summary of sputtering parameters (U = 2 kV, p_{total} = 2.5 × 10^{-2} mbar for all thin films. Single-side polished (SSP), double-side polished (DSP). * fully closed valve.

Nr.	Thin Film	p_{H2} (10^{-4} mbar)	p_{H2} % of Total Pressure (%)	Sputtering Time (min)	Flow (sccm)	Substrate
R1	a-SiN$_x$	0	0	30	0 *	SSP
R2	a-SiN$_x$	0	0	80	0 *	SSP, DSP glass
S1	a-SiN$_x$:H	0.5	0.2	30	0.9	SSP
S2	a-SiN$_x$:H	0.8	0.32	30	1.6	SSP
S3	a-SiN$_x$:H	1.5	0.6	30	3	SSP
S4	a-SiN$_x$:H	3.3	1.32	30	6	SSP
S5	a-SiN$_x$:H	7.9	3.16	30	12	SSP
S6	a-SiN$_x$:H	3.3	1.32	80	12	SSP, DSP glass

Single-side and double-side polished (SSP and DSP, respectively) intrinsic un-doped crystalline (001) Si wafers and ~40 mm × 40 mm size soda lime glass slides with nominal thickness of 4 mm (Guardian Orosháza Ltd., Orosháza, Hungary) were used as substrates. The target–substrate distance was constantly kept as 50 mm. The sputtering process was applied at room temperature. The thin film properties are not directly determined by the process parameters because the plasma diagnostic results are not available. Fourier-transform infrared spectroscopy (FTIR) was used for investigating the hydrogen bond configuration. FTIR measurements were performed by a Varian 7000 FTIR spectrometer (Agilent Technologies, Santa Clara, CA, USA) connected to a UMA 600 IR microscope equipped with a mercury–cadmium–telluride (MCT) detector. The absorbance spectra were recorded in the wavenumber range between 600 and 4000 cm^{-1} with spectral resolution of 4 cm^{-1}. The evaluation of the measured spectra including the baseline correction by adjusted polynomial fit was done using Origin 2019b (64-bit) software, version 9.6.5.169. Transmission electron microscopy (TEM, Philips CM20 with 200kV accelerating voltage, Hillsboro, OR, USA) and Cs-corrected (S)TEM (FEI Themis 200 with accelerating voltage 200 kV) were applied for the structural characterization of the thin films. TEM samples were prepared by conventional Ar ion milling technique. Scanning electron microscopy

(SEM, LEO 1540 XB with accelerating voltage 5 kV) was used to investigate the morphology of the thin films' surfaces before and after annealing. SE measurements were performed by a Woollam M2000DI UV–VIS ellipsometer (Woollam Co., Lincoln, NE, USA) at the angles of 70° and 75° with a compensator frequency of 20 Hz. The wavelengths used for the measurements ranged from 200 to 1600 nm. The hydrogen-free SiN_x thin film was modeled by the Cauchy–Urbach equation and it was considered as a mixture of silicon (Si), hydrogen-free stoichiometric silicon nitride (Si_3N_4), and void. Based on this consideration, the thin films were modeled by the Bruggeman-type EMA model. All modeling and calculations were performed using Woollam VASE software (versions of 3.83 and 3.84). The activation energy of the surface modification was determined from the Arrhenius equation using an optical measurement configuration, the details of which are given elsewhere [22]. The 1.6 MeV Rutherford backscattering spectrometry/elastic recoil detection analysis (RBS/ERDA) measurements were performed in a scattering chamber with a two-axis goniometer, which was connected to the 5 MV EG-2R Van de Graaff accelerator operated at the Wigner Research Center of Physics in Budapest, Hungary. The $4He^+$ analyzing ion beam was collimated using two sets of four-sector slits. The width × height of the beam spot was 0.2 mm × 1 mm. The beam divergence was kept below 0.06°. The beam current was measured using a transmission Faraday cup. The vacuum in the scattering chamber was kept at about 10^{-4} Pa. Hydrocarbon deposition was avoided by liquid N_2-cooled traps along the beam path and around the wall of the chamber. A Mylar foil with thickness of 6 µm was placed before the window of the ERDA detector to capture backscattered He^+ ions. Kapton foil as reference was used for calibration of the hydrogen content of the samples. ORTEC Si surface barrier detectors mounted at scattering angles of $\Theta = 165°$ and 20° were used to detect RBS and ERDA spectra, respectively. The detector resolution was about ~20 keV. The spectra were measured at sample tilt angles of 7° and 80° for RBS, and 80° for ERDA.

3. Results and Discussion

3.1. Optical Characterization of Thin Films

The refractive index and extinction coefficient are related to the interaction between a thin film and the incident light, indicating the optical properties of the thin films. SE is one of the most popular tools for characterizing the optical properties of different materials. The polarization state of the light changes while it is reflected or transmitted by the sample. Detection and interpretation of this change are the basis of the ellipsometric method. The thin films, reference R1 and thin films with hydrogen addition (S1–S6), were measured by SE in reflectance mode, where the complex reflectance ratio ($\bar{\rho}$) is recorded by the instrument. This quantity is usually expressed by the ψ and Δ ellipsometric angles in the following form:

$$\bar{\rho} = \frac{\bar{r}_p}{\bar{r}_s} = tg\psi \times e^{i\Delta} \qquad (1)$$

where $\bar{\rho}$, $\bar{r}_p$, and $\bar{r}_s$ refer to the complex reflectance ratio, reflectance coefficient in p (parallel to the incident plane), and s (perpendicular to the incident plane) directions, respectively. The evaluation of the measurement data includes the modeling and fitting procedure. The differences between the modeled (generated) and real (measured) ψ and Δ ellipsometric angles are minimized by adjusting the free parameters of the applied model. The effects of the hydrogen flow applied to the different optical properties of the a-SiN_x:H thin films are shown in Figure 2. The refractive index of 1.96 is characteristic for the reference thin film (R1) sputtered in nitrogen atmosphere (Figure 2a). The addition of hydrogen until 3 sccm gives no or minimal effect on the refractive index. The increasing of hydrogen flow from 6 to 12 sccm results in a decrease in the refractive index from 1.96 to 1.89 (Figure 2a). We note that K. Mokkedem et al. reported a refractive index of 1.8 for hydrogenated silicon nitride thin films deposited under similar conditions using DC magnetron sputtering [16].

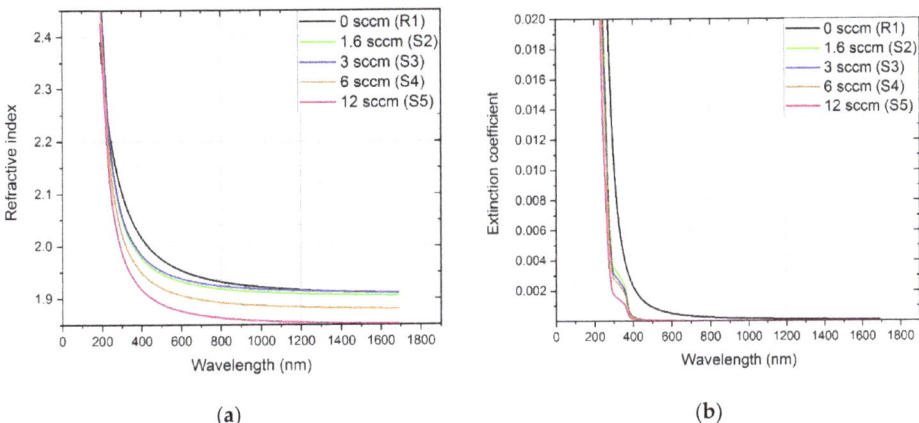

Figure 2. Effect of hydrogen flow on the optical properties of a-SiN$_x$ thin films: (**a**) refractive index, (**b**) extinction coefficient.

The refractive index of stoichiometric SiN is greatly dependent upon the deposition conditions, but it is greater than 2.0 at 630 nm [23]. The lower values of our RF-sputtered thin films indicate the presence of the non-stoichiometric SiN$_x$ phase. The extraction of the optical constants using the optical reflection spectra alone is usually very infrequent and complicated. The extinction coefficient values exhibit a large difference between the thin films grown with or without hydrogen (Figure 2b). There is a shift in the extinction coefficient from 400 to 300 nm, when hydrogen/nitrogen reactive sputtering is used.

The refractive index measured at 550 nm wavelength has a decreasing character as the partial pressure of hydrogen is increased during the sputtering process (Figure 3). K. Mokkedem et al. showed that the refractive index of their films vs. partial pressure of H$_2$ had a similar character. They confirmed the effect of the hydrogen partial pressure on the refractive index. These variations may be explained by hydrogen and nitrogen incorporation into the thin films [16].

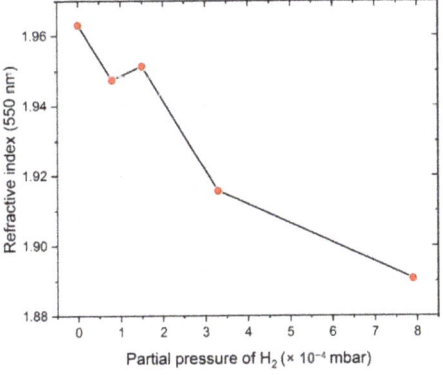

Figure 3. Refractive index of hydrogenated silicon nitride (SiN$_x$:H) thin films at a wavelength of 550 nm as a function of H$_2$ partial pressure.

3.2. Bonding Configuration and Chemical Composition

Understanding the chemical properties of silicon nitride and the role played by the hydrogen atoms that are incorporated during the growth of the film is a key factor for a-SiN$_x$:H applicability. FTIR is a widely used method for characterization of the bonding configuration in thin films. FTIR measurements allow to characterize the bonds involved in the material, and eventually to determine the bond densities and thus hydrogen concentration thanks to the method developed by Lanford and Rand [24]. In the case of a-SiN$_x$:H, the presence of different hydrogen bonds can also be assumed based on the Lanford–Rand method [24].

Absorbance spectra of a-SiN$_x$ (R2) and a-SiN$_x$:H (S6) revealed the bonding configuration of the investigated thin films (Figure 4), consisting of typical absorption bands for different hydrogen bonding to nitrogen and silicon. The peak at 880 cm^{-1} is associated with the Si–N stretching mode [25]. Peaks observed at 1175, 2200, and 3335 cm^{-1} refer to the presence of the N–H bending mode, Si–H stretching mode, and N–H stretching mode, respectively [26].

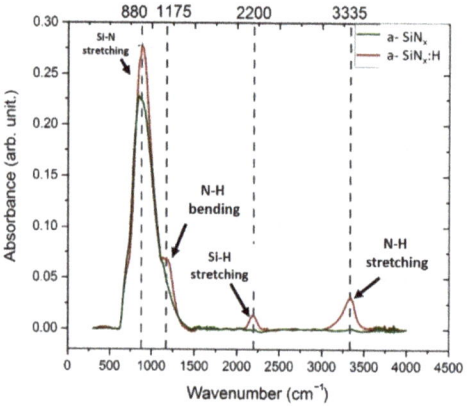

Figure 4. Absorbance spectra of reference a-SiN$_x$ (R2) and a-SiN$_x$:H (S6) thin films.

The concentrations of the N–H and Si–H bonds (C_{Y-H}) were calculated using the Lanford–Rand method [24] defined in the following form:

$$C_{Y-H} = \frac{A_{Y-H}}{ln10 \times \sigma_{Y-H}} = \frac{\int_{(Y-H)^s}^{(Y-H)^f} \alpha(\omega)d(\omega)}{ln10 \times \sigma_{Y-H}} \quad (2)$$

where Y corresponds to silicon (Si) or nitrogen (N) atoms, A_{Y-H}, σ_{Y-H}, $(Y-H)^s$, $(Y-H)^f$, $\alpha(\omega)$ are the concentration of Y–H bond in the cm^{-3} unit, the normalized absorption area of the Y–H band, the absorption cross-section of the Y–H bond in the cm^2 unit, the beginning wavenumber of the Y–H band in the cm^{-1} unit, the final wavelength of the Y–H band in the cm^{-1} unit, and the spectral absorption coefficient, respectively.

The σ_{N-H} and σ_{Si-H} absorption cross-sections were determined by Lanford and Rand [24]: $\sigma_{N-H} = 5.3 \times 10^{-18}$ cm^2 and $\sigma_{Si-H} = 7.4 \times 10^{-18}$ cm^2. These values were used in Equation (2) above. Table 2 summarizes the parameters and the results of the calculations for both Si–H and N–H bond concentrations.

Table 2. Calculation parameters and results of the Si–H and N–H bond concentrations.

Bond	$(Y-H)^s$ (cm^{-1})	$(Y-H)^f$ (cm^{-1})	A_{Y-H}	σ_{Y-H} (10^{-18} cm^{-2})	C_{Y-H} (10^{20} cm^{-3})
N–H	3026	3502	0.066	5.3	3.49
Si–H	1994	2299	0.016	7.4	0.61

If the hydrogen is one coordinated, then the total concentration of bonded hydrogen atoms in the film can be calculated by the sum of the concentration of N–H and Si–H bonds:

$$C_{H,\,bound} = C_{N-H} + C_{Si-H} \quad (3)$$

Based on Equation (3), the amount of total bonded hydrogen content of the S6 sample is 4.1×10^{20} at/cm^3.

Rutherford backscattering spectrometry (RBS) in combination with elastic recoil detection analysis (ERDA) measurements were performed to obtain thin film elemental concentration depth profiles.

The atomic concentrations of silicon, nitrogen, free (unbounded), and bounded hydrogen as well as thin film atomic densities were determined (Table 3) from measured RBS/ERDA spectra, considering the layer thicknesses determined from TEM measurements. The samples show comparable atomic densities; only a-SiN$_x$:H sputtered with high hydrogen flow (S4) showed a lower value of ~4.9×10^{22} at/cm^3. In the case of the thin (S1) and thick a-SiN$_x$:H (S6) samples, the atomic density of measured hydrogen was found to be 0.46×10^{22} and 0.59×10^{22} at/cm^3, respectively (Table 3). We note that RBS/ERDA is not sensitive to the chemical bonding states and measures all the hydrogen content even if it is in atomic, bounded, or in H$_2$ molecular form.

Table 3. Atomic layer densities and concentrations (at.%) and atomic densities (at/cm^3) for the Si, N, and H components of a-SiN$_x$:H layers as evaluated from RBS/ERDA measurements using layer thicknesses (in nm) obtained from TEM analysis. * fully closed valve.

H$_2$ Flow (sccm)	Atomic Layer Density (10^{22} at/cm^3)	Silicon (Si)		Nitrogen (N)		Hydrogen (H)	
		at.%	(10^{22} at/cm^3)	at.%	(10^{22} at/cm^3)	at.%	(10^{22} at/cm^3)
0 *	6.8	43.2	3.1	50	3.59	6.8	0.46
0.9	6.25	40.4	2.48	52	3.19	7.6	0.47
1.6	6.6	43.3	2.73	45.3	2.85	11.4	0.75
3	4.9	35.2	1.72	52.4	2.57	12.4	0.61
12	6	33.1	1.99	57	3.42	9.9	0.59

Lanford et al. noted that the exact values of hydrogen content cannot be measured by FTIR, but a good approximation of the bonding structure is represented [24]. In contrast to effusion measurements, FTIR shows only the bonded amount of hydrogen in the material, whereas the effusion measurement detects the total hydrogen content including atomic hydrogen as well.

However, effusion of hydrogen is mostly detected as molecular H$_2$. In RF-sputtered a-SiN$_x$:H (S6), the concentration of bonded hydrogen calculated from FTIR spectra is only 4 at.%. The ERDA measurement confirmed 10 at.% of the total hydrogen concentration (Table 3). This means that 6 at.% hydrogen was incorporated during growth in molecular form. This fact is in good agreement with density measurements (Table 3). According to Dekkers et al., a low-density material induces the release of hydrogen in molecular form (H$_2$) and a denser SiN$_x$ makes the hydrogen desorption slower but its atomic form is preferred [27].

3.3. Activation Energy

Owing to the annealing chemical reactions such as breaking up and rebuilding of different bonds, diffusion of hydrogen atoms or molecules could take place. The energy must be provided for a certain chemical reaction (activation energy) of the process. As a result of chemical reactions, the annealing quality of the layer surface can be changed. The estimated activation energy can be detected by this change. The arrangement of the measurement described in Ref. [22] enables the determination of onset time of the layer surface change by monitoring the elapsed time and the alteration of the reflectivity of the sample surface.

The a-SiN$_x$:H thin film sputtered at a H$_2$ flow of 6 sccm (S4) was investigated at different temperatures. The temperature and the corresponding onset time values are summarized in Table 4. The Arrhenius equation [28] describes the rate constant of a chemical reaction as a function of the temperature in terms of two empirical factors, namely E_{exp} experimental activation energy and A pre-exponential factor:

$$r(T) = Ae^{-\frac{E_{exp}}{k_B T}} \tag{4}$$

where r, T, and k_B refer to the rate constant, temperature in K units, and Boltzmann constant, respectively. A and E_{exp} can be determined from the parameters (increment and slope) of linear fit on $\ln(k)$ against $1/T$ experimental points.

Table 4. Measurement data of the a-SiN$_x$:H thin film sputtered at a H$_2$ flow of 6 sccm (S4) for the Arrhenius plot.

Temperature (K)	Onset Time (s)
320	292
323	149
326	57
330	28

The r reaction rate constant against the inverse temperature of annealing and the linear fit on the data points are shown in Figure 5. The experimental activation energy (E_{exp}) deduced from the slope of the fitted line is $E_{exp} = 2.19 \pm 0.17$ eV. It must be noted that the error of E_{exp} is much bigger than the error of the linear fit due to the uncertainty of time and temperature measurement.

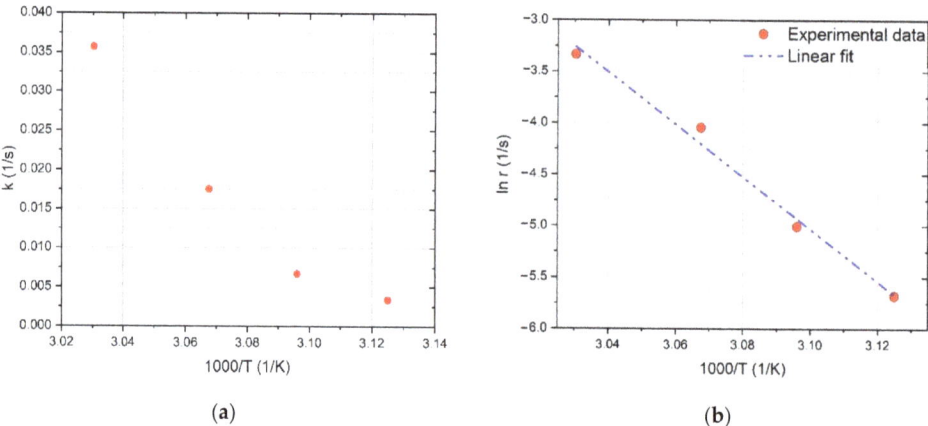

Figure 5. Arrhenius plot of the a-SiN$_x$:H thin film sputtered at a H$_2$ flow of 6 sccm (S4): (**a**) experimental data, (**b**) linear fit.

3.4. Structural and Morphological Characterization

Structural investigations confirmed the film thickness between ~142 nm and 155 nm (Figure 6). The selected area electron diffractions (SAEDs) proved that the deposited layer structure is amorphous without any crystalline structure (Figure 6). The SAED was provided by the ProcessDiffraction program [29–31].

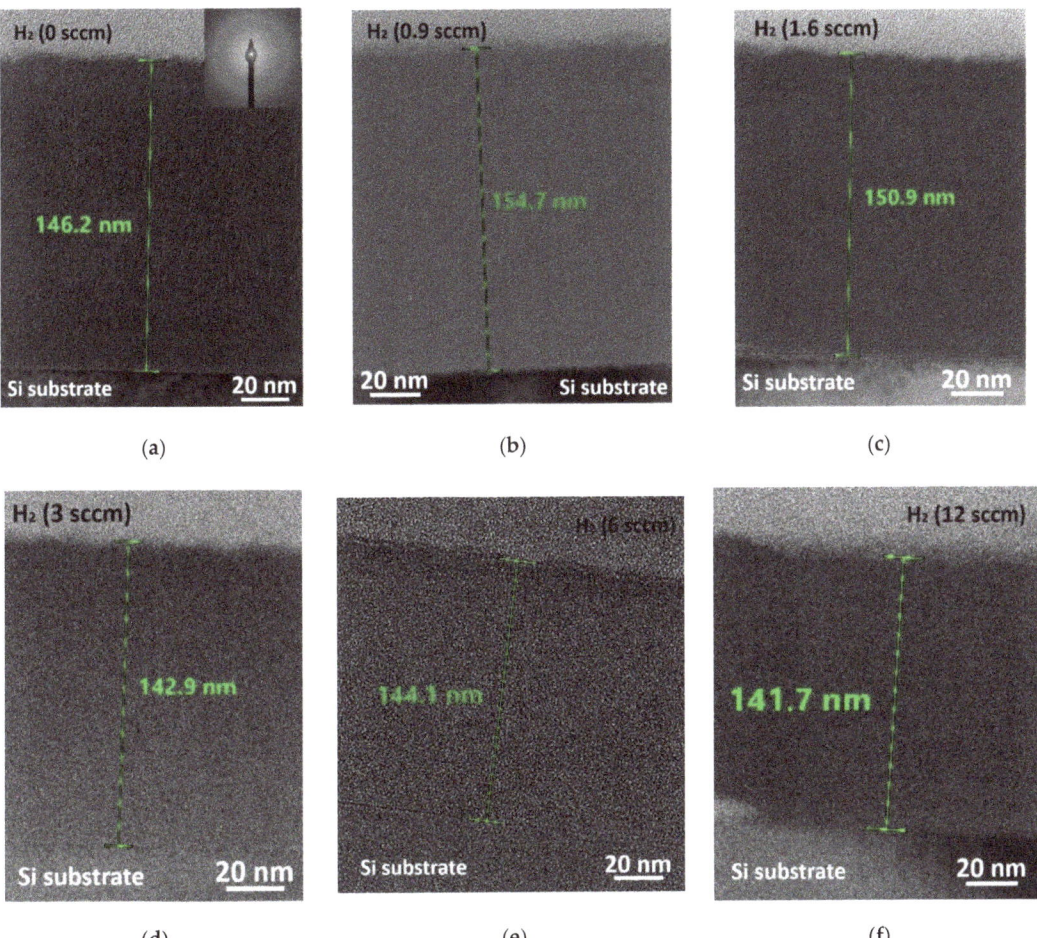

Figure 6. Cross-section TEM images of a-SiN and a SiN$_x$:H thin films sputtered at different H$_2$ flow: (**a**) 0 sccm (R1) with selected area electron diffraction (SAED) detail, (**b**) 0.9 sccm (S1), (**c**) 1.6 sccm (S2), (**d**) 3 sccm (S3), (**e**) 6 sccm (S4), and (**f**)12 sccm (S5).

Such a large difference indicates a more complex incorporation mechanism of hydrogen in the silicon–nitrogen network and the presence of voids in a-SiN$_x$:H thin films. These structural observations correlated with the refractive index values. The refractive index decreased with increasing hydrogen flow and porosities during the growth process of the thin films. Similar results were obtained by AJ. Flewitt et al. when the nitrogen incorporation into a-Si:N:H films was caused by the dissociation of NH$_3$ molecules, leading to the reaction with a growing surface. This process resulted in a lower refractive index of the a-SiN$_x$:H [32].

TEM provides direct information on the structure of materials. Furthermore, the high-angle annular dark-field (HAADF) STEM allows the visibility of the porosities. The detailed study confirmed the dense thin film (Figure 7a) during hydrogen-free sputtering and the porous structure with nanometer-scale porosities homogenously distributed in the thin film sputtered at 1.6 sccm hydrogen flow (Figure 7b).

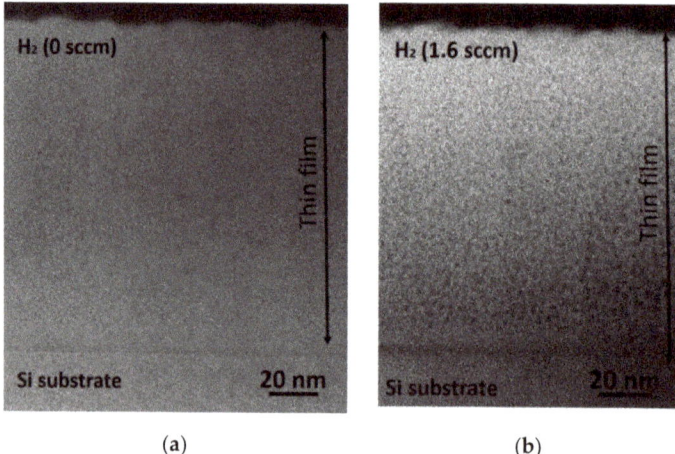

Figure 7. High-angle annular dark-field (HAADF) STEM images of a-SiN$_x$ thin films: (**a**) hydrogen-free a-SiN$_x$ thin film (S1), (**b**) a-SiN$_x$:H (S3).

The out-diffusion of hydrogen due to annealing plays a prominent role in the densification of thin films. The temperature of 800 °C was considered as a critical temperature, since the a-SiN$_x$ thin films are often subjected to thermal processes including rapid thermal annealing [33]. Morphological investigations of the thin film's surface before (Figure 8a) and after (Figure 8b) annealing show that the surface of a-SiN$_x$:H thin film changes due to heat treatment (Figure 8). The hydrogen, because of its molecular form, is released at a temperature of ~65 °C from the film. Blisters with a diameter of the order of 100 nm are created on the surface of the thin films. It is known from the literature [10,34,35] that, upon annealing, both hydrogen and nitrogen releases from a-SiN$_x$:H thin film were observed.

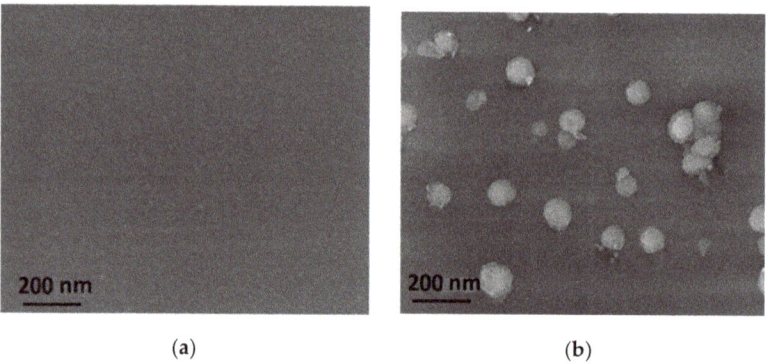

Figure 8. SEM images of a-SiN$_x$ sputtered at a H$_2$ flow rate of 12 sccm: (**a**) surface before annealing, (**b**) surface after annealing at 800 °C.

Similar to the effect reported in Ref. [22], blisters on the layer surface could be burst bubbles filled with hydrogen and/or nitrogen containing molecules. Initially, the volume of these bubbles was increasing due to thermal expansion of the fill-up gas during heat treatment. Then, at a critical point (at a given temperature after a certain time), the bubbles burst, leading to blister creation on the surface.

4. Conclusions

Amorphous hydrogenated silicon nitride thin films (a-SiN$_x$:H) have widespread applications from device passivation to light emitting diodes and antireflective coatings for solar cells. Chemical vapor deposition (CVD) and physical vapor deposition (PVD) are the most common techniques for silicon nitride film deposition with or without hydrogen addition. CVD-deposited film always contains hydrogen and its amount cannot be controlled directly during the preparation process. Due to this fact, the alternative fabrication method for controlled hydrogen concentration in a direct way from zero hydrogen content by adjusting the applied hydrogen gas flow to the chamber could be PVD (e.g., RF sputtering).

In this work, a-SiN$_x$:H films were sputtered at various H$_2$ flows with average thickness of 150 nm, and the effect of hydrogen incorporation on structural and optical properties was studied. The detailed structural characterization confirmed the formation of a dense thin film at hydrogen-free sputtering and a porous structure with homogenously distributed nanometer-scale porosities caused by hydrogen addition. The refractive index of 1.96 was characteristic for hydrogen-free SiN$_x$ thin films. Hydrogen flows up to 3 sccm were found to have no or minimal effect on the refractive index; for flows from 6 to 12 sccm, the refractive index decreased from 1.96 to 1.89, which can be explained by the hydrogen and nitrogen incorporation into the thin films. The calculations from FTIR spectra showed that a-SiN$_x$:H sputtered at 6 sccm H$_2$ flow presented a concentration of bounded hydrogen of ~4 at.%. The ERDA measurements confirmed a total hydrogen content of 10 at.%. This means that 6 at.% hydrogen was incorporated in a molecular form during the layer growth, which explained the lower density of the thin films. The out-diffusion of hydrogen due to annealing plays a prominent role in the densification of thin films. The molecular form of hydrogen is released at a temperature of ~65 °C from the film. Blisters with 100 nm diameter are created on the surface of the thin films. The low activation energy calculated by the Arrhenius method refers to significant diffusion of hydrogen molecules.

Author Contributions: N.H.: sample deposition, optical characterization, writing of the manuscript; R.L.: RF sputtering; M.S.: Arrhenius calculations; Z.Z.: ERDA measurements; P.P.: ellipsometry measurements; J.M.: FITR measurements; Z.F.: TEM measurements; C.B.: supervising, writing of the manuscript; K.B.: supervising, structural correlations, writing. All authors have read and agreed to the published version of the manuscript.

Funding: This research was funded by OTKA grant Nr. K131515, OTKA grant Nr. K131594, FLAG-ERA NKFIH 127723, NKFIH-NNE 129976.

Institutional Review Board Statement: Not applicable.

Informed Consent Statement: Not applicable.

Data Availability Statement: The data presented in this study are available on request from the corresponding author.

Acknowledgments: The authors would like to thank Levente Illés for SEM measurements and Andrea Fenyvesiné Jakab from the Centre for Energy Research for TEM sample preparations.

Conflicts of Interest: The authors declare no conflict of interest.

References

1. Liao, W.; Zeng, X.; Wen, X.; Chen, X.; Wang, W. Annealing and excitation dependent photoluminescence of silicon rich silicon nitride films with silicon quantum dots. *Vacuum* **2015**, *12*, 147–151. [CrossRef]
2. Kim, D.W.; Kim, D.H.; Kim, H.J.; So, H.W.; Hong, M.P. Effect of ammonia (NH$_3$) plasma treatment on silicon nitride (SiN$_x$) gate dielectric for organic thin film transistor with soluble organic semiconductor. *Curr. Appl. Phys.* **2011**, *11*, S67–S72. [CrossRef]
3. Jin, Z.; Jung, S.; Duy, N.V.; Hwang, S.; Jang, K.; Lee, K.; Lee, J.; Hyungjun, P.; Kim, J.; Son, H.; et al. Characterization of MONOS nonvolatile memory by solid phase crystallization on glass. *Surface Coat. Technol.* **2008**, *202*, 5637–5640. [CrossRef]
4. Swatowska, B.; Stapinski, T. Amorphous hydrogenated silicon-nitride films for applications in solar cells. *Vacuum* **2008**, *82*, 942–946. [CrossRef]
5. Qiu, W.; Kang, Y.M.; Goddard, L.L. Quasicontinuous refractive index tailoring of SiN$_x$ and SiOxNy for broadband antireflective coatings. *Appl. Phys. Lett.* **2010**, *96*, 14116. [CrossRef]

6. Duerinck, F.; Szlufcik, J. Defect passivation of industrial multicrystalline solar cells based on PECVD silicon nitride. *Sol. Energy Mater. Sol. Cells* **2002**, *72*, 231–246. [CrossRef]
7. Boehme, C.; Lucovsky, G. H loss mechanism during anneal of silicon nitride: Chemical dissociation. *J. Appl. Phys.* **2000**, *88*, 6055. [CrossRef]
8. Bommali, R.K.; Ghosh, S.; Khan, S.A.; Srivastava, P. Hydrogen loss and its improved retention in hydrogen plasma treated a-SiN$_x$:H films: ERDA study with 100 MeV Ag^{7+} ions. *Nucl. Instrum. Methods Phys. Res. B* **2018**, *423*, 16–21. [CrossRef]
9. Guler, I. Optical and structural characterization of silicon nitride thin films deposited by PECVD. *Mater. Sci. Eng. B* **2019**, *246*, 21–26. [CrossRef]
10. Santos-Filho, P.; Stevens, G.; Lu, Z.; Koh, K.; Lucovsky, G. *Hydrogen Release and Si-N Bond-Healing Infrared Study of Rapid Thermal Annealed Amorphous Silicon Nitride Thin Films*; Cambridge University Press: Cambridge, UK, 1995.
11. Martil, I.; del Prado, A.; San Andres, E.; Gonzalez Daz, G.; Martnez, F.L. Rapid thermally annealed plasma deposited SiN$_x$:H thin films: Application to metal-insulator-semiconductor structures with Si, In$_{0.53}$Ga$_{0.47}$As, and InP. *J. Appl. Phys.* **2003**, *94*, 2642. [CrossRef]
12. Martinez, F.L.; Ruiz-Merino, R.; del Prado, A.; San Andres, E.; Martil, I.; Gonzalez-Diaz, G.; Jeynes, C.; Barradas, N.P.; Wang, L.; Reehal, H.S. Bonding structure and hydrogen content in silicon nitride thin films deposited by electron cyclotron resonance plasma method. *Thin Solid Film.* **2004**, *459*, 203–207. [CrossRef]
13. Verlaan, V.; van der Werf, C.H.M.; Houweling, Z.S.; Romijn, I.G.; Weeber, A.W.; Dekkers, H.F.W.; Goldbach, H.D.; Schropp, R.E.I. Multi-crystalline Si solar cells with very fast deposited (180 nm/min) passivating hot-wire CVD silicon nitride as antireflection coating. *Prog. Photovolt. Res. Appl.* **2007**, *15*, 563–573. [CrossRef]
14. Kessels, W.M.M.; Hong, J.; van Assche, F.J.H.; Moschner, J.D.; Lauinger, T.; Soppe, W.J.; Weeber, A.W.; Schram, D.C.; van de Sanden, M.C.M. High-rate deposition of a-SiNx:H for photovoltaic application y the expanding thermal plasma. *J. Vac. Sci. Technol. A* **2002**, *20*, 1704. [CrossRef]
15. Signore, M.A.; Sytchkova, A.; Dimaio, D.; Cappello, A.; Rizzo, A. Deposition of silicon nitride thin films by RF magnetron sputtering: A material and growth process study. *Opt. Mater.* **2012**, *34*, 632–638. [CrossRef]
16. Mokeddem, K.; Aoucher, M.; Smail, T. Hydrogenated amorphous silicon nitride deposited by DC magnetron sputtering. *Superlattices Microstruct.* **2006**, *40*, 598–602. [CrossRef]
17. Banerjee, R.; Bandyopadhyay, A.K.; Rath, A.K.; Batablyal, A.K.; Barua, A.K. Properties of hydrogenated amorphous Si-N films prepared by r.f. magnetron sputtering with emphasis on the non stochiometric region. *Thin Solid Film.* **1990**, *192*, 295–307. [CrossRef]
18. Tiron, V.; Velicu, I.; Pana, I.; Cristea, D.; Rusu, B.G.; Dinca, P.; Porosnicu, C.; Grigore, E.; Munteanu, D.; Tascu, S. HiPIMS deposition of silicon nitride for solar cell application. *Surf. Coat. Technol.* **2018**, *344*, 197–203. [CrossRef]
19. Boulesbaa, M. Optical and Physico-chemical Properties of Hydrogenated Silicon Nitride Thin Films—Effect of the Thermal Annealing. *Spectrosc. Lett.* **2016**, *50*, 5–12. [CrossRef]
20. de Brito Mota, F.; Justo, J.F.; Fazzio, A. Hydrogen role on the properties of amorphous silicon nitride. *J. Appl. Phys.* **1999**, *86*, 1843–1847. [CrossRef]
21. Tao, S.X.; Theulings, A.M.M.G.; Prodanovic, V.; Smedley, J.; van der Graaf, H. Optical Properties of Silicon-Rich Silicon Nitride (Si$_x$N$_y$H$_z$) from First Principles. *Computation* **2015**, *3*, 657–669. [CrossRef]
22. Serényi, M.; Csík, A.; Hámori, A.; Kalas, B.; Lukács, I.; Zolnai, Z.S.; Frigeri, C. Diffusion and reaction kinetics governing surface blistering in radio frequency sputtered hydrogenated a-Si$_x$Ge$_{1-x}$ (0 < x < 1) thin films. *Thin Solid Film.* **2019**, *679*, 58–63.
23. Jellison, G.E.; Modine, F.A.; Doshi, P.; Rohatgi, A. Spectroscopic ellipsometry characterization of thin-film silicon nitride. *Thin Solid Film.* **1998**, *313–314*, 193–197. [CrossRef]
24. Lanford, W.A.; Rand, M.J. The hydrogen content of plamsa-deposited silicon nitride. *J. Appl. Phys.* **1978**, *49*, 2473. [CrossRef]
25. Parsons, N.; Souk, J.H.; Batey, J. Low hydrogen content stoichiometric silicon nitride films deposited by plamaenhanced chemical vapor deposition. *J. Appl. Phys.* **1991**, *70*, 1553–1560. [CrossRef]
26. Rostaing, J.C.; Cros, Y.; Gujrathi, S.C.; Poulain, S. Quantitative infrared characterization of plasma enhanced CVD silicon oxynitride films. *J. Non-Cryst. Solids* **1987**, *97–98*, 1051–1054. [CrossRef]
27. Dekkers, H.F.W.; Carnel, L.; Beaucarne, G.; Beyer, W. Diffusion mechanism of hydrogen through PECVD SiN$_x$:H for a fast defect passivation of mc-Si solar cells. In Proceedings of the 20th European Photovoltaic Solar Energy Conference, Barcelona, Spain, 6–10 June 2005.
28. Menzinger, M.; Wolfgang, R. The Meaning and Use of the Arrhenius Activation Energy. *Angew. Chem.* **1969**, *8*, 438–444. [CrossRef]
29. Lábár, J.L. Electron Diffraction Based Analysis of Phase Fractions and Texture in Nanocrystalline Thin Films, Part I: Principles. *Microsc. Microanal.* **2008**, *14*, 287–295. [CrossRef]
30. Lábár, J.L. Electron Diffraction Based Analysis of Phase Fractions and Texture in Nanocrystalline Thin Films, Part II: Implementation. *Microsc. Microanal.* **2009**, *15*, 20–29. [CrossRef]
31. Lábár, J.L. Electron Diffraction Based Analysis of Phase Fractions and Texture in Nanocrystalline Thin Films, Part III: Application Examples. *Microsc. Microanal.* **2012**, *18*, 406–420. [CrossRef]
32. Flewitt, A.J.; Dyson, A.P.; Robertson, J.; Milne, W.I. Low temperature growth of silicon nitride by electron cyclotron resonance plasma enhanced chemical vapour deposition. *Thin Solid Film.* **2001**, *383*, 172–177. [CrossRef]

33. Jafari, S.; Hirsch, J.; Lausch, Z.; John, M.; Bernhard, N.; Meyer, S. Composition Limited Hydrogen Effusion Rate of a-SiN$_x$:H Passivation Stack. *Aip Conf. Proc.* **2019**, *2147*, 050004.
34. Fitzner, M.; Abelson, J.R.; Kanicki, J. Investigation of hydrogen and nitrogen thermal stability in PECVD a-SiN$_x$:H. *MRS Online Proc. Libr. Arch.* **2011**, *258*, 362–368. [CrossRef]
35. Martnez, F.L.; del Prado, A.; Martil, I.; Gonzalez-Daz, G.; Selle, B.; Sieber, I. Thermally induced changes in the optical properties of SiN$_x$:H films deposited by the electron cyclotron resonance plasma method. *J. Appl. Phys.* **1999**, *86*, 2055. [CrossRef]

Article

The Study of the Influence of Matrix, Size, Rotation Angle, and Magnetic Field on the Isothermal Entropy, and the Néel Phase Transition Temperature of Fe_2O_3 Nanocomposite Thin Films by the Monte-Carlo Simulation Method

Dung Nguyen Trong [1,*], Van Cao Long [1] and Ştefan Ţălu [2]

[1] Institute of Physics, University of Zielona Góra, Prof. Szafrana 4a, 65-516 Zielona Góra, Poland; vancaolong2020@gmail.com
[2] The Directorate of Research, Development and Innovation Management (DMCDI), Technical University of Cluj-Napoca, 15 Constantin Daicoviciu St., 400020 Cluj-Napoca, Romania; stefan_ta@yahoo.com
* Correspondence: dungntdt2018@gmail.com

Citation: Nguyen Trong, D.; Cao Long, V.; Ţălu, Ş. The Study of the Influence of Matrix, Size, Rotation Angle, and Magnetic Field on the Isothermal Entropy, and the Néel Phase Transition Temperature of Fe_2O_3 Nanocomposite Thin Films by the Monte-Carlo Simulation Method. *Coatings* **2021**, *11*, 1209. https://doi.org/10.3390/coatings11101209

Academic Editors: Angela De Bonis and Devis Bellucci

Received: 12 August 2021
Accepted: 29 September 2021
Published: 2 October 2021

Publisher's Note: MDPI stays neutral with regard to jurisdictional claims in published maps and institutional affiliations.

Copyright: © 2021 by the authors. Licensee MDPI, Basel, Switzerland. This article is an open access article distributed under the terms and conditions of the Creative Commons Attribution (CC BY) license (https://creativecommons.org/licenses/by/4.0/).

Abstract: In this paper, the study of the influence of the matrix structure (mxm) of thin-film, rotation angle (α), magnetic field (B), and size (D) of Fe_2O_3 nanoparticle on the magnetic characteristic quantities such as the magnetization oriented z-direction (M_{zE}), z-axis magnetization (M_z), total magnetization (M_{tot}), and total entropy (S_{tot}) of Fe_2O_3 nanocomposites by Monte-Carlo (MC) simulation method are studied. The applied MC Metropolis code achieves stability very quickly, so that after 30 Monte Carlo steps (MCs), the change of obtained results is negligible, but for certainty, 84 MCs have been performed. The obtained results show that when the mxm and α increase, the magnetic phase transition appears with a very small increase in temperature Néel (T_{Ntot}). When B and D increase, T_{Ntot} increases very strongly. The results also show that in Fe_2O_3 thin films, T_{Ntot} is always smaller than with Fe_2O_3 nano and Fe_2O_3 bulk. When the nanoparticle size is increased to nearly 12 nm, then T_{Ntot} = T = 300 K, and between TNtot and D, there is a linear relationship: T_{Ntot} = −440.6 + 83D. This is a very useful result that can be applied in magnetic devices and in biomedical applications.

Keywords: isothermal entropy; Néel phase transition temperature; synthetic nanomaterials; Fe_2O_3 thin films; the Monte-Carlo method

1. Introduction

Thin films, two-dimensional physical systems with different structures and materials, have been studied for many decades, especially after the discovery of the so-called giant magnetoresistance effect [1,2] and topological phase transitions [3,4]. Research in this domain not only gives us theoretical results concerning the foundations of modern physics but also leads to the nanotechnology of thin layers, which gives different new functional materials with significant applications in practice. Recently, Prof. Mirosław Dudek et al. [5–9] intensively studied mechanical and magnetic characteristics of different two-dimensional nanocomposite materials. We will discuss their results obtained in detail below. We would like to emphasize that our paper follows this direction of study.

Nowadays, magnetic nanocomposites play an important role in science and technology [10,11]. It is important in practical applications to find materials that can be used in devices such as force sensors and refrigeration equipment. These devices always ensure requirements such as high load capacity, good corrosion resistance, and lightweight. To ensure these requirements, we choose Fe_2O_3 nano synthetic thin film as the subject of this paper. The reason is that the synthetic nano-Fe_2O_3 thin film is an antiferromagnetic material [12–17] with many advantages relative to the original material. Moreover, it is widely used in practical applications such as recording equipment [18,19], refrigeration equipment [20,21], printing devices [22], photocatalyst materials [23,24], ion recovery [25], magnetic nanofilm [26], magnetic resonance [27,28], magnetic fluids [29], pigments [30], gas sensors [31,32], the biomedical field [33–37], and spintronics [38,39]. In addition, Fe_2O_3 is also a material with many structures, such as α-Fe_2O_3 (hematite), β-Fe_2O_3 and γ-Fe_2O_3 (maghemite), and ε-Fe_2O_3 [40]. Among them, the α-Fe_2O_3 hematite structure is the most commonly used today, while the ε-Fe_2O_3 structure is the most difficult structure to be manufactured. The ε-Fe_2O_3 structure has not been studied extensively. Applying the experimental method, Zuohui Cheng et al. [41] have successfully determined the influence of enthalpy and entropy on the size of Fe_2O_3 materials. They concluded that when the size (D) increases from D = 19.3 nm to D = 140.5 nm, the total magnetization (M_{tot}) increases, and the Néel phase transition temperature (T_{Ntot}) increases from T_{Ntot} = 742 K to T_{Ntot} = 897 K [41]. Liu et al. [42] proposed an approximate expression for the enthalpy and entropy of the nanoparticles without considering the change in temperature during the transition. Cui et al. [43] demonstrated that the crystallinity densities of the two transitions are equal. Zhang et al. [44] successfully determined the Néel phase transition temperature of Fe_2O_3 when considered system transitioned from the maghemite phase to the α hematite phase, namely T_{Ntot} = 623 K at D = 4 nm and T_{Ntot} = 723 K at D = 24 nm. Mendili et al. [45], Wenger et al. [46] have successfully determined that the magnetization (M) of γ-Fe_2O_3 nanoparticles increases when D increases (from D = 2 nm to D = 4 nm). Hou et al. [47] suggested that the coercivity of nano-Fe_2O_3 increases when the D size increases (from D = 12 nm to D = 40 nm). Jun Wang et al. [48] successfully determined the T_{Ntot} of α-Fe_2O_3 as T_{Ntot} = 930 K. In the study of José Luis García-Muñoz [49] for ε-Fe_2O_3, this temperature was T_{Ntot} = 850 K. The obtained results show that, for bulk or nanomaterials, an increase in D leads to an increase in T_{Ntot}. Additionally, it was found that, for Fe_2O_3 synthetic thin film at room temperature T = 300 K, there exists a very large magnetic force of approximately 20 kOe. Due to this, Fe_2O_3 materials are used frequently in high-density magnetic recording devices, high-frequency electromagnetic waves absorbers [50–52]. On the other hand, ε-Fe_2O_3 is a ferromagnetic material [53]. In some kind, it is an intermediate form between γ-Fe_2O_3 and α-Fe_2O_3, so its Neel phase transition temperature ranges from T_{Ntot} = 150 K to T_{Ntot} = 500 K. It has a lowest temperature of T_{Ntot} = 150 K [54,55], and it is characterized by a decrease in coercivity and saturation magnetization [54]. In addition, at the cooling condition of Fe_2O_3, the T_{Ntot} value is in the range of T_{Ntot} = 85–150 K, when there is a change in spins in the rhombohedral structure and a non-monotonic change in the Fe-O bond [49,56], even though the Curie temperature (T_{Ntot}) of ε-Fe_2O_3 has a value of $T_{Ntot} \sim 500$ K [57–60]. Additionally, magnetic ε-Fe_2O_3 nanoparticles can also be obtained by vibrating magnetometer, which shows T_{Ntot} > 500 K and a maximum value of T_{Ntot} = 850 K. The simulation method has received great attention, because researchers can study materials in extreme conditions, such as at high temperature (T), T = 700 K, pressure (P), P = 360 GPa, and with atomic size below 2 nm, where the experimental methods cannot be applied [61–63]. Meanwhile, the magnetic refrigeration technology industry attracts great attention from researchers. Cooling technology is based on the magnetic effect of materials, which was first discovered by E. Warburg in 1881 [64]. When the material is placed in the magnetic field (B), under the action of the magnetic field, the spins rotate in the direction of the magnetic field, leading to the total magnetic (M_{tot}) and total entropy (S_{tot}) change (increase or decrease), depending on the nature of magnetic materials. In the field of refrigeration technology at room temperature T = 300 K, researchers often use soft magnetic materials which operate at room temperatures, such as Gadolinium with T_{Ntot} = 296 K. It is the first used material with very high cost, poor oxidation resistance, and low magnetic value. For that, researchers try to find materials with high oxidation resistance and low cost, which exhibit high magnetic properties that are essential for future applications. Presently, materials such as transition metals or rare earth materials are still considered materials with great potential for refrigeration applications at room temperature [65–71]. When materials with high or low magnetism are considered, researchers concentrate on the entropy change (ΔS). This quantity is determined through the Maxwell relationship and the isothermal magnetization curve. However, up to now, the use of the Maxwell relationship to determine entropy change is still causing much controversy and discussions about it [72–76]. Giguère et al. [72] successfully determined the entropy change based on the Maxwell relationship. Balli et al. [75] and Liu et al. [76] also successfully demonstrated that the Maxwell relationship no longer exists when the material is near the phase transition (paramagnetic phase, ferromagnetic phase, antiferromagnetic). They suggested that the cause of the appearance of the giant magnetic field at room temperature is due to the morphological leading to changes in the magnetic transition temperature at the T~110 K. In addition, there is an arrangement of the magnets. Different cations lead to a structural change from tetrahedral to octahedral, driven by spin-oriented ions at the magnetic particles, and a magnetic transition occurs at T_{Ntot}~150 K. T. Muto et al. [77] studied the entropy, whereas P. Fratzl [78] and Dieter [79] determined the phase transition diagram. In 2010, Jiri Tucek et al. [80] successfully determined the existence of the huge magnetic field of nano ε-Fe_2O_3 at room temperature. Recently, M.R. Dudek et al. have successfully determined the giant magnetic field of Fe_3O_4 nanomaterials at room temperature [6], the magnetic domain in auxetic materials [7,8], and successfully constructed a single-body Hamiltonian function [6,12,17]. Combining with the phase space average field in [9] to study the magnetism of Fe_3O_4 nanoparticles [81], Dung et al. [82] determined the magnetic properties of Fe nanomaterials by the Monte-Carlo simulation method with the classical Heisenberg model. In addition, researchers also successfully studied the influence of factors such as temperature, number atomic, pressure, annealing time on structure, electronic structure, phase transition, and crystallization progress of material metals [83–87], alloys [88–92], oxide [61–63], and polymers [93,94]. Here, a question appears: how to determine the magnetic characteristic quantities of materials such as magnetization in all directions and entropy of the material when the size of the material is less than 10 nm. To answer this question, in this article, we focus on a study of the influence of factors such as material size (mxm), magnetic field (B), and nanoparticle size on magnetization in the direction priority z-axis (M_{zE}), z-axis magnetization (M_z), total composite magnetization (M_{tot}), and total entropy of Fe_2O_3 nano synthetic thin films. For this purpose, we used the Monte-Carlo simulation method. The obtained results will serve as a basis for future experimental studies when we try to apply Fe_2O_3 nano synthetic thin films to smart devices and refrigeration equipment.

2. Method of Calculation

Initially, the two-dimensional model for Fe_2O_3 nano synthetic thin film is constructed by creating a 2D matrix square (mxm). These matrices are composed of nonmagnetic squares and linked together by hinges defined as the intersection points between the corners of the squares (red color in the figure), and these 2D square matrices may be deformed [95]. Then, spherical Fe_2O_3 nanoparticles were put into the 2D matrix square. For simplicity, we treat each nanoparticle as a magnetic spin. When the thin film model is not deformed, the rotation angle of these 2D square matrices has the value $\alpha = 0°$ (Figure 1a). When the 2D square matrices rotate, the corresponding rotation angle varies from $\alpha = 0°$ to $\alpha = 90°$ (Figure 1b).

In this model, the size of each 2D square is the diameter of the inserted Fe_2O_3 nanoparticle, with $D = 2R$, where R is the radius of the nanoparticle. We studied the magnetic properties of Fe_2O_3 nano synthetic thin films by applying a potential force field to the nano synthetic thin film, with the value of the Hamilton function of the form (Equation (1)) [6], and numerical simulation was performed by the Monte-Carlo method.

$$H_{mfa}^{(i)} = -K_a V \cos^2 \alpha_i - BM\cos\alpha_i - \sum_j K_{ij} \vec{m_i} \cdot \langle \vec{m_j} \rangle \quad (1)$$

where $K_a = \frac{1}{k_B T} \cdot \frac{\mu_0}{4\pi M^3}$, $V = \frac{4}{3}\pi a_0^3$, a_0 = 8.394 Å, and $K_{ij} = \frac{\mu_0 M^2}{4\pi d^3}$, $d = \sqrt{2} a \sin(\alpha + \frac{\pi}{4})$, $a = D + a_0$, $D = 2R_g$.

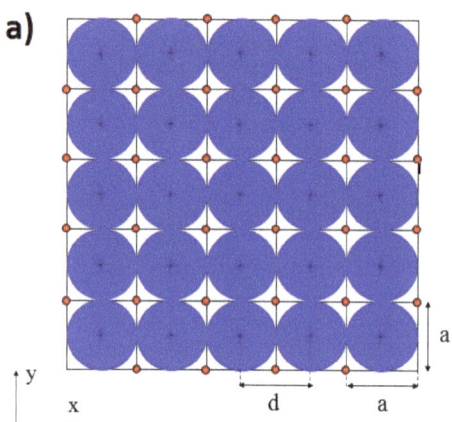 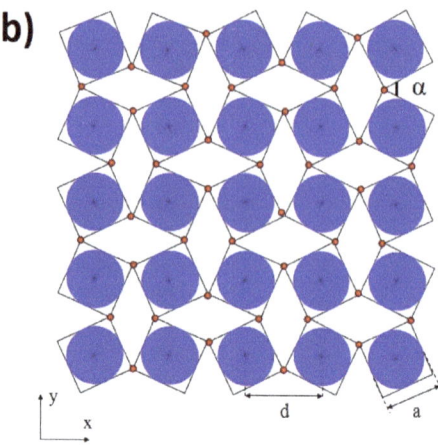

Figure 1. The initial shape of materials has a square structure with a rotation angle $\alpha = 0°$ (**a**), rotation angle $\alpha \neq 0°$ (**b**).

In it, K_a is the uniaxial magnetic anisotropy energy; the magnetic coefficient is $\mu_0 = 4\pi \times 10^{-7}$ m/A; V is the volume of the nanoparticle; a_0 is the lattice constant; K_{ij} is the interaction energy between the ith nanoparticle and the nearest j^{th} nanoparticle; and Boltzmann's constant $k_B = 1.38 \times 10^{-23}$ J/K $= 8.617 \times 10^{-5}$ eV/K, where B is the magnetic field, M is magnetic moment, α_i is the rotation angle of the ith 2D square matrix, m_i and m_j are the magnetic moments of the ith and j^{th} atom, and d, a, R_g, and D are the distance between the centers of the two nearest nanoparticles, the size of the square edge, the displacement radius, and the size of the Fe_2O_3 nanoparticle, respectively. Conversely, the size of the model is determined by the following formula (Equation (2)):

$$L = (m - 1) \times d \quad (2)$$

Here, m is the number of rows (columns) of the matrix. The interaction between the Fe_2O_3 nanoparticles was determined by the magnetic dipole interaction. Then, Fe_2O_3 nanoparticles are affected by magnetic moments in all directions.

In the spherical coordinate system: $m_x = \sin\alpha\cos\varphi$, $m_y = \sin\alpha\sin\varphi$, $m_z = \cos\alpha$; $0 < \alpha < 180°$, $0 < \varphi < 360°$. The total magnetic moment (M_{tot}) is expressed by the following expression (Equation (3)) [6]:

$$M_{tot} = \sqrt{M_x^2 + M_y^2 + M_z^2} = \pm 1, \quad M_{tot} = \frac{1}{N}\sum_{i=1}^{N} S_i, \quad S_i = \pm 1, \quad (3)$$

where $S_i = +1$ with spin up, and $S_i = -1$ with spin down of Fe_2O_3 nanoparticles.

In the nano synthetic thin film with the size L, $V_s = b(Ld)^2$ is the thin film body and b, d, α, and φ are, respectively, the nanoparticle thickness, the distance between the nanoparticles, the polarization angle (pointing the direction of the magnetic moment in the x–y plane), and the azimuth (pointing the direction of the magnetic moment for the z-axis).

To study the magnetic properties of Fe_2O_3 nano synthesized thin films, various authors have applied the Monte Carlo method in numerical simulations [96–98]. To increase the accuracy of the results, we used periodic boundary conditions to eliminate surface effects. The obtained results are also compared with the results of the density functional theory method to increase the accuracy of the obtained results.

Ising's 2D model is placed in the magnetic field B = 0.1 T while the influencing factors, such as model size (mxm), rotation angle (α), and the external magnetic field (B), are changed. To simulate numerically, we used the Metropolis algorithm in the framework of the Monte-Carlo method and surveyed magnetic characteristic quantities in temperatures from T = 0 to 600 K, with a total number of MC simulation steps 5×10^4 corresponding to 84 MC steps for each temperature T = 1 K. It has been emphasized in [6] that the Monte-Carlo Metropolis code becomes stable very quickly. It follows from Figure 3c of this paper that, from the vicinity of room temperature to the larger temperatures, the results obtained after 20 Monte Carlo steps (MCS) are practically the same as those after 200 MCS. Our calculations show that, after 30 MCS, the change in obtained results is negligible, but for certainty, 84 MCS have been performed.

The simulation method is based on a random generation of energy variation of the system. Next, we rotated their magnetic moment from $\vec{m} = (m_x, m_y, m_z)$ to $\vec{m}' = (m'_x, m'_y, m'_z)$ and calculated the energy values H_m, H_m' with the corresponding probability distribution (Equation (4)) [6]:

$$P(E) = \frac{\exp(-\beta \Delta E)}{Z}, \ Z = \sum_{i}^{N} \exp(-\beta \Delta E) \text{ and } \Delta E = H_m - H_m', \qquad (4)$$

where P(E) is the probability value of finding spin min (1, $\exp(-\beta \Delta E)$) in a state; $\beta = 1/k_B T$; T is the temperature, Z is the partition function, ΔE is energy variation of the system generated randomly.

To analyze the model, we calculate the magnetic characteristic quantities of the considered system, such as the total entropy (Equation (5)) [6] of the Fe_2O_3 nano synthetic thin film with the following expression:

$$\frac{S_{tot}}{Nk_B} = \frac{1}{Nk_B} \sum_i S_i \qquad (5)$$

To determine the Néel phase transition temperature (T_{Ntot}) of Fe_2O_3 nano synthetic thin films, the intersection between the magnetization curve with the entropy curve is fixed. The entire numerical simulation was carried out based on the Python programming code provided by Prof. M.R. Dudek [6]. This code was properly modified for our purpose and was applied on the computational server system of the Institute of Physics, Department of Physics and Astronomy, Zielona Gora University, Poland.

3. Results and Discussion

3.1. Magnetic Characteristic Quantities

We determine the magnetic characteristic quantities of Fe_2O_3 nano synthesized thin films, such as the preferred magnetization in the z-axis (M_{zE}), the magnetization in the z-axis (M_z), the total magnetization (M_{tot}), total entropy (S_{tot}), and Néel phase transition temperature (T_{Ntot}).

The Néel phase transition temperature is the phase transition temperature of a material from an antiferromagnetic state to a superparamagnetic state. To determine the characteristic quantities, as has been emphasized above, we treat each spherical nanoparticle as a spin (with D = 6 nm and the magnetic moment is determined by Equation (3)). The results are shown in Figure 2.

The results show that when the Fe_2O_3 nano synthetic thin film is placed in the magnetic field (B), B = 0.1 Tesla (T), and the spin of the nanoparticles is rotated by an angle $\alpha = 90°$, the shape of the synthesized thin film nano Fe_2O_3 with mxm = 5×5, nano size (D), D = 6 nm corresponding to the size L = 27 nm is given in Figure 2a. The relationship between the magnetization oriented in the direction of the z-axis (M_{zE}) is shown by the black line in Figure 2b; magnetization in the z-axis (M_z) is shown by the green line in Figure 2c. Synthetic magnetization of Fe_2O_3 materials is given by dark blue line in Figure 2d and synthetic entropy (S_{tot}) is drawn in red color when the temperature increases.

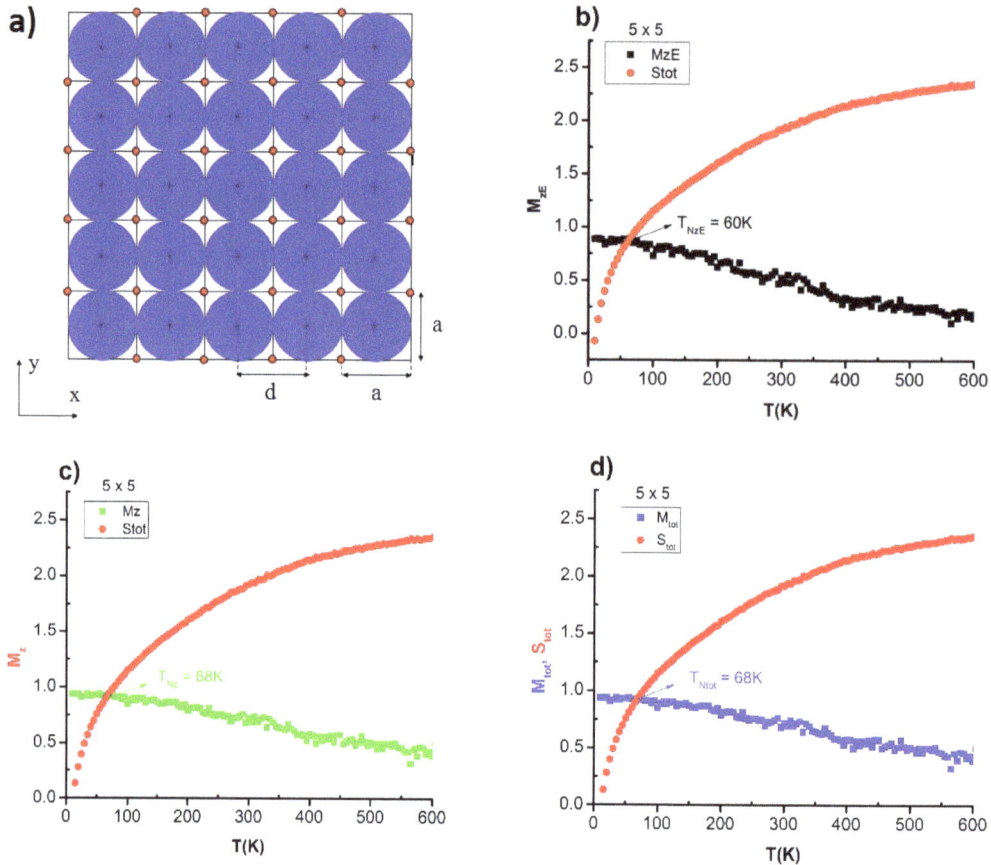

Figure 2. Shape (**a**), the relationship between magnetic characteristic quantities with different temperatures of Fe_2O_3 nanocomposite thin film with mxm = 5 × 5 (L = 27 nm), nano size (D), D = 6 nm, magnetic field B = 0.1 T, and rotation angle α = 90°. The magnetization in the preferred z-axis (**b**), the magnetization in the z-axis (**c**), and the total magnetization (**d**) with the total entropy.

Dudek et al. [6] successfully determined the Néel phase transition temperature of the Fe_3O_4 nano synthetic thin film and showed that the cause of this phenomenon is due to the magnetic effect of the spins. For this reason, we omit the determination of magnetic characteristic quantities such as magnetization (M), specific heat (C_v), magnetic susceptibility (χ), and energy (E) of the thin film synthesized Fe_2O_3 nano and only focus our attention on studying the relationship between the characteristics of the magnetization M (M_{zE}, M_z, M_{tot}) with the total entropy (S_{tot}) when the temperature (T) increases (what it has been demonstrated above in Figure 2). The total entropy is determined according to the formula (5). At T = 10 K, M_{zE} = 0.886, M_z = 0.941, M_{tot} = 0.941, S_{tot} = −0.069, when T increases from T = 10 K to T = 600 K with the temperature shift dT = 5 K, all values of magnetization M (M_{zE}, M_z, M_{tot}) decrease as M_{zE} decreases from M_{zE} = 0.886 to M_{zE} = 0.156, M_z decreases from M_z = 0.941 to M_z = 0.390, and M_{tot} decreases from M_{tot} = 0.941 to M_{tot} = 0.394, which leads to an increase in S_{tot} from S_{tot} = −0.069 to S_{tot} = 2.354. The displacement of the spins corresponds to the probability of finding the existence of spins in a given state at a certain temperature. The lines of the magnetization M_{zE}, M_z, M_{tot}, and the total entropy S_{tot} intersect at a point, which is called the magnetic phase transition point or the Néel phase transition temperature (T_{Ntot}). The intersection between M_{zE} and S_{tot} is T_{NzE} = 60 K;

between M_z and S_{tot} is $T_{Nz} = 68$ K; and between M_{tot} and S_{tot} is $T_{Ntot} = 68$ K. This is the Néel phase transition temperature from the antiferromagnetic state to the superparamagnetic state. This result is completely consistent with the magnetic effect results previously obtained with Fe_3O_4 nano synthetic thin films at room temperature [6]. To confirm that, we study the factors affecting the isotherm entropy and Néel temperature of Fe_2O_3 nano synthesized thin films with D = 6 nm.

3.2. The Influence of Different Factors
3.2.1. Effect of the Synthetic Thin-Films
Effect of the Synthetic Thin Film Size

To study the effect (mxm) of Fe_2O_3 nano synthetic thin films, the size model increases from mxm = 5 × 5 (L = 27 nm) to mxm = 10 × 10 (L = 62 nm), 15 × 15 (L = 96 nm), 20 × 20 (L = 130 nm), 30 × 30 (L = 198 nm), 40 × 40 (L = 267 nm) with D = 6 nm. The considered system is placed in a magnetic field (B), B = 0.1 T with a rotation angle of α = 90°. The result obtained is shown in Figure 3.

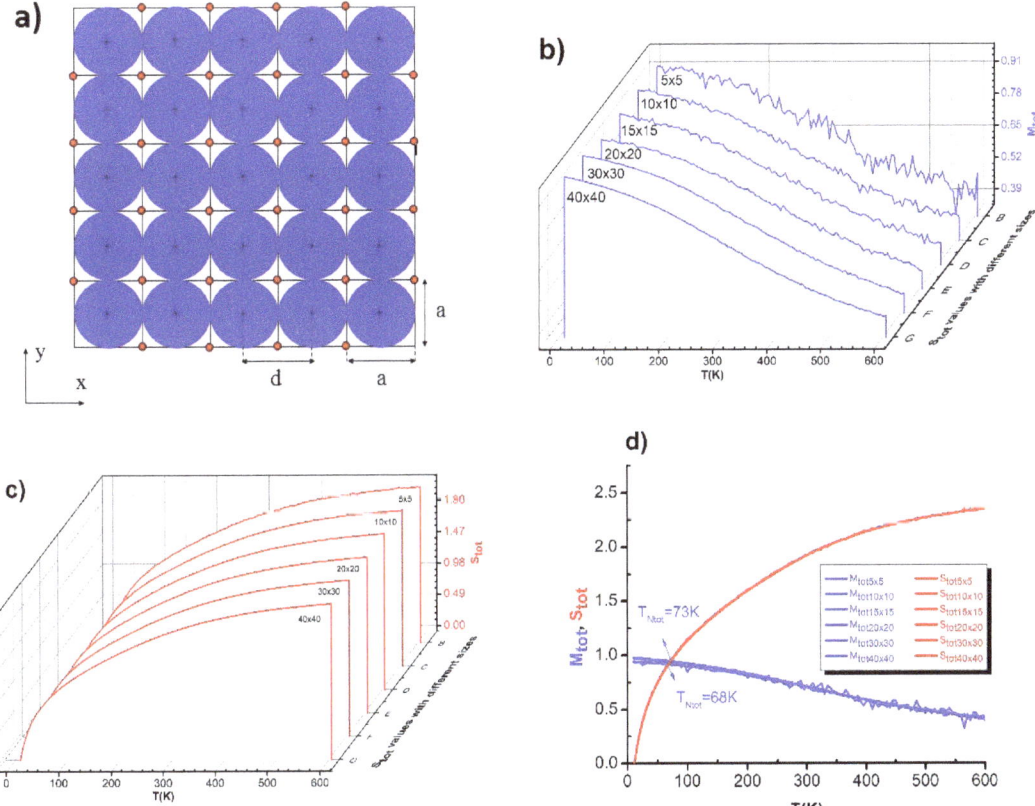

Figure 3. Shape (**a**), synthetic magnetization (**b**), synthetic entropy (**c**), Néel phase transition temperature (**d**) of Fe_2O_3 nano synthetic thin film in magnetic field B = 0.1 T with a rotation angle of α = 90° for different matrices.

The obtained results show that the Fe_2O_3 nano synthetic thin film of size L = 27 nm has the shape given in Figure 3a. The total magnetization (M_{tot}) is shown by the blue line in Figure 3b, and the total entropy (S_{tot}) is shown by the red line in Figure 3c. When temperature (T) increases from T = 10 K to T = 600 K, the magnetization (M_{tot}) decreases from $M_{tot} = 0.941$ to $M_{tot} = 0.394$, S_{tot} increases from $S_{tot} = -0.069$ to $S_{tot} = 2.354$, and

Neel's magnetic phase transition temperature (T_{Ntot}) increases slightly from $T_{Ntot} = 68$ K to $T_{Ntot} = 68, 68, 68, 71, 73$ K (Figure 3d). The reason for these changes is that increasing temperature T leads to a shift of the domain walls. When the thin film size increases from L = 27 nm to L = 62, 96, 130, 198, 267 nm, the M_{tot} increases slightly from $M_{tot} = 0.941$ to $M_{tot} = 0.944, 0.947, 0.952, 0.972, 0.983$, respectively, because the increase in the thin film size of L leads to an increase in the density of spins. The obtained results are completely consistent with the simulation results of amorphous Fe nanoparticles [82]. The cause of this phenomenon is due to the size effect (when increasing the lattice size L leads to an increase in T_{Ntot}) with a negligible increase in results (about 9%). We chose Fe_2O_3 nano synthetic thin film with the nano size D = 6 nm, mxm = 20 × 20 corresponding to the size L = 130 nm as standard to study other influencing factors. Further, we investigated the influence of the spin angle of spin on the magnetic characteristic quantities.

Effect of the Spins Rotation Angle

Similarly, as in the case of analyzing the effect of synthetic thin film size, we consider the influence of spin angle using Fe_2O_3 nano synthetic thin film with L = 130 nm (D = 6 nm), B = 0.1 T, with an angle α that changes from α = 0° to α = 90°. The obtained results are shown in Figure 4.

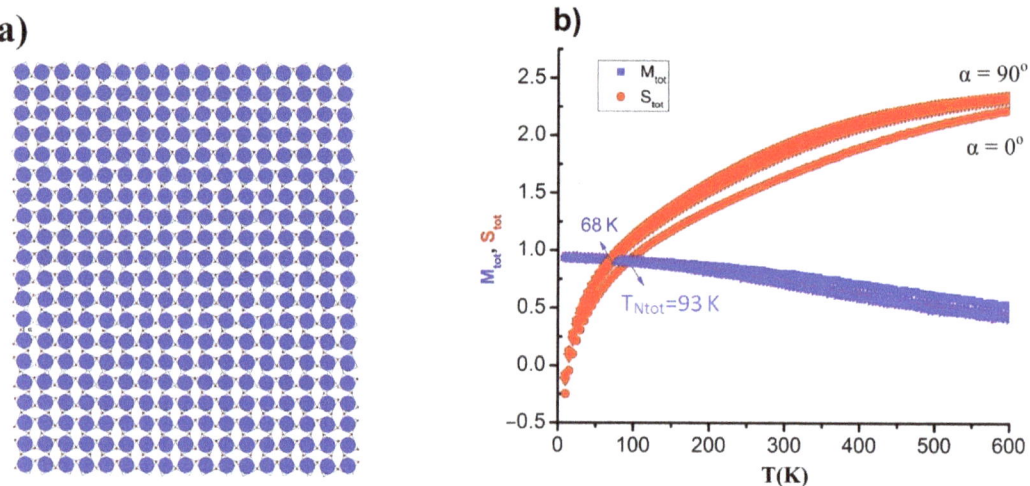

Figure 4. Shape (**a**), Néel phase transition temperature (**b**) of Fe_2O_3 nano synthetic thin film in magnetic field B = 0.1 T, L = 130 nm for different rotation angles.

The results show that when the Fe_2O_3 nano synthetic thin film of size L = 130 nm is placed in a magnetic field (B) with B = 0.1 T and the rotation angle (α) increases from α = 0° to α = 30°, 45°, 60°, 90°, the shape of the thin film is shown as in Figure 4a. The M_{tot} increases slightly from $M_{tot} = 0.939$ to $M_{tot} = 0.940, 0.940, 0.941, 0.941$, because an increase in the size L leads to an increase in the density of spins, whereas S_{tot} increases slightly from $S_{tot} = -0.249$ to $S_{tot} = -0.133, -0.107, -0.088, -0.076$. This leads to the decrease in magnetic phase transition temperature (T_{Ntot}) from $T_{Ntot} = 93$ K to $T_{Ntot} = 86, 75, 70, 68$ K (Figure 4b). The cause of the change in T_{Ntot} is the fact that an increase in the rotation angle leads to an increase in spin spacing (d) and to a decrease in the magnetization M_{tot}. The distance between the nanoparticle centers increases d > a, and in consequence, total entropy S_{tot} decreases. It follows from obtained results that when L = 27 nm increases to L = 62, 96, 130, 198, 267 nm, the T_{Ntot} increases from 68 K to 75 K, and when the rotation angle increases from α = 0° to α = 90°, T_{Ntot} decreases from $T_{Ntot} = 93$ K to $T_{Ntot} = 68$ K with L = 130 nm. The increase in size leads only to an insignificant change of T_{Ntot}, and the rotation angle of the matrix will be a very convenient parameter for experimental studies with different types of materials used to manufacture thin films. Through the research results on the influence of thin films on the magnetic properties of Fe_2O_3 nano synthetic thin films, we conclude that the influence factor of the thin film is very small, almost negligible. So, a question arises: what causes the increase or decrease in T_{Ntot}? To study the influencing factors of Fe_2O_3 nanoparticles, we chose a thin film with a size L = 130 nm with a rotation angle of α = 90°. To answer this question, we continued the study of the influence of nanoparticles and the impact factors of the external magnetic field on experiments.

3.2.2. Effect of Fe_2O_3 Nanoparticles

To study the influence of Fe_2O_3 nanoparticles, we used again a matrix of size L = 130 nm, D = 6 nm with a rotation angle of the matrix α = 90°.

Effect of the External Magnetic Field

Let us consider the Fe_2O_3 nano synthetic thin film with nanoparticle size D = 6 nm, L = 130 nm into the external magnetic field B with different intensities. The obtained results are shown in Figure 5.

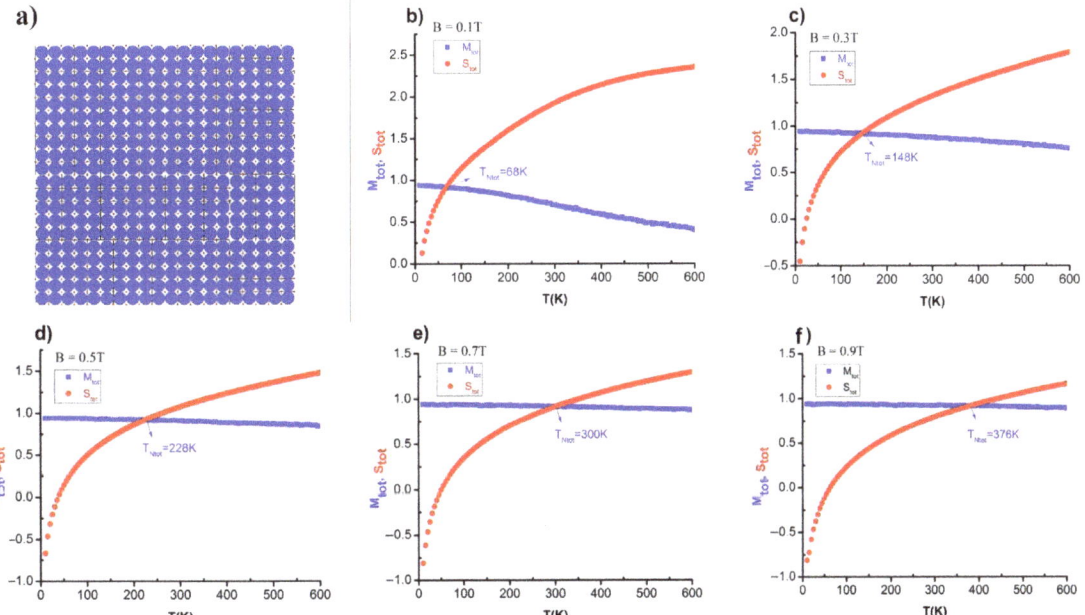

Figure 5. Shape (**a**), Néel phase transition temperature of Fe_2O_3 nano synthetic thin film with L = 130 nm, rotation angle α = 90° for different magnetic fields B = 0.1 T (**b**), B = 0.3 T (**c**), B = 0.5 T (**d**), B = 0.7 T (**e**), and B = 0.9 T (**f**).

The results show that when Fe_2O_3 nano synthetic thin film with matrix size L = 130 nm, nanoparticle size D = 6 nm is placed in a magnetic field (B) with B = 0.1 T, the shape of Fe_2O_3 thin film is as in Figure 5a. The M_{tot} composite magnetization decreases from M_{tot} = 0.941 to M_{tot} = 0.404, the S_{tot} composite entropy increases from S_{tot} = −0.076 to S_{tot} = 2.353, and the magnetic phase transition temperature is T_{Ntot} = 68 K (Figure 5b). When the external magnetic field increases from B = 0.1 T to B = 0.3, 0.5, 0.7, 0.9 T, the point above of total magnetization M_{tot} increases slightly from M_{tot} = 0.941 to M_{tot} = 0.943, the total entropy always increases from S_{tot} = −0.076 to S_{tot} = −0.932, the lower point of M_{tot} increases again from M_{tot} = 0.404 to M_{tot} = 0.754, 0.844, 0.881, 0.889, and S_{tot} again decreases from S_{tot} = 2.353, 1.789, 1.479, 1.295, 1.165. This leads to a decrease in the magnitude of M, while the magnitude of S_{tot} increases. It implies that S_{tot} tends to shift towards the negative axis, which leads to a corresponding increase in T_{Ntot}: $T_{Ntot} = 68$ K at B = 0.1 T (Figure 5b), $T_{Ntot} = 148$ K at B = 0.3 T (Figure 5c), $T_{Ntot} = 228$ K at B = 0.5 T (Figure 5d), $T_{Ntot} = 300$ K at B = 0.7 T (Figure 5e), $T_{Ntot} = 376$K at B = 0.9 T (Figure 5f). The cause of the change in T_{Ntot} is that an increase in B leads to the stronger orientation of the spins in the preferred direction of the magnetic field and they rotate very strongly with a large magnetic field. So, there is another problem: how to increase T_{Ntot} with a small external magnetic field?

Effect of Nanoparticle Size

When nanoparticle size (D) increases, we obtain the results shown in Figure 6.

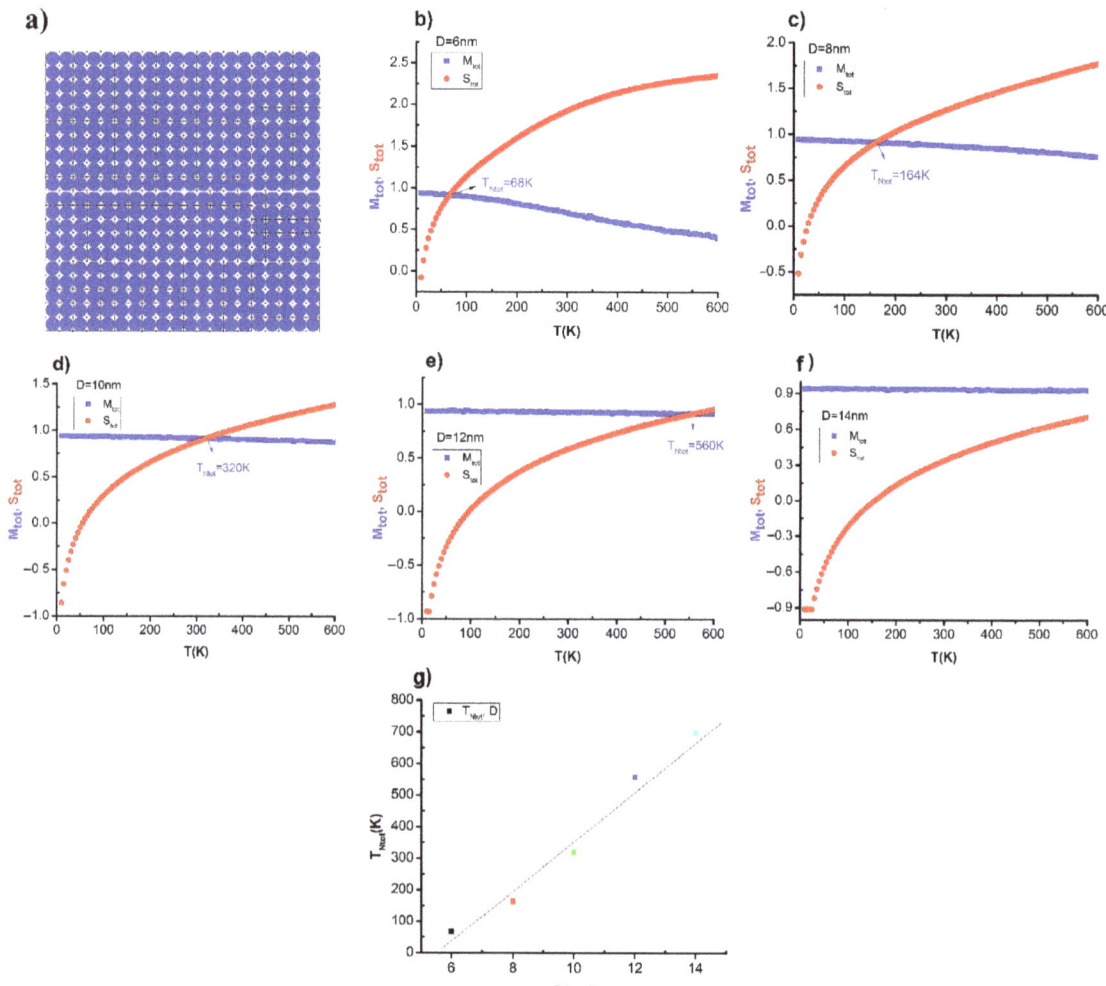

Figure 6. The Néel phase transition temperature (T_{Ntot}) of Fe_2O_3 nano synthetic thin film with mxm = 20 × 20, L = 130 nm, B = 0.1 T, rotation angle $\alpha = 90°$ for different nanoparticle sizes: the shape (**a**), the results for the size D = 6 nm (**b**), D = 8 nm (**c**), D = 10 nm (**d**), D = 12 nm (**e**), D = 14 nm (**f**) and T_{Ntot} depends on D (**g**).

The results show that when Fe_2O_3 nano synthetic thin film with nanoparticle size D = 6 nm (L = 130 nm) is placed in a magnetic field (B), with B = 0.1 T, rotation angle $\alpha = 90°$, the shape of the film is as in Figure 6a. The M_{tot} composite magnetization decreases from $M_{tot} = 0.941$ to $M_{tot} = 0.404$, S_{tot} composite entropy increases from $S_{tot} = -0.076$ to

S_{tot} = 2.353. The T_{Ntot} magnetic phase transition temperature is T_{Ntot} = 68 K (Figure 6b). When the size increases from D = 6 nm (L = 130 nm) to D = 8, 10, 12, 14 nm (L = 168, 206, 244, 282 nm), then upper point of total magnetization M_{tot} increases slightly from M_{tot} = 0.941 to M_{tot} = 0.983, the total entropy always increases from Stot = −0.076 to Stot = −0.083, the lower point of M_{tot} increases again from M_{tot} = 0.404 to Mtot = 0.876, 0.929, 0.940, 0.935, and S_{tot} again decreases from S_{tot} = 2.353, 1.281, 0.705, 0.310, 0.003, which leads to the decrease in magnitude of M_{tot}, whereas the magnitude of S_{tot} increases. S_{tot} tends to shift towards the negative axis, leading to an increase in T_{Ntot}: T_{Ntot} = 68 K at D = 6 nm, (L = 130 nm) (Figure 6b), T_{Ntot} = 164 K at D = 8 nm, (L = 168 nm) (Figure 6c), T_{Ntot} = 320 K at D = 10 nm, (L = 206 nm) (Figure 6d), T_{Ntot} = 560 K at D = 12 nm, (L = 244 nm) (Figure 6e), $T_{Ntot} \geq$ 600 K at D = 14 nm, (L = 282 nm) (Figure 6f). The M_{tot} curve and S_{tot} curve intersect at a point called the point of magnetic phase transition (T_{Ntot}). Thereby, it shows that there is a relationship between D and T_{Ntot}: for D = 6 nm (L = 130 nm), T_{Ntot} = 68 K; for D = 8 nm (L = 168 nm), T_{Ntot} = 164 K; for D = 10 nm (L = 206 nm), T_{Ntot} = 320 K; when D = 12 nm (L = 244 nm), T_{Ntot} = 560 K; and for D = 14 nm (L = 282 nm), T_{Ntot} > 600 K. This result is completely consistent with the results obtained in [99] for the magnetic phase transition temperature (T_{Ntot}) of Fe_3O_4 nanoparticles (as D increases, the T_{Ntot} increases and reaches a maximum value T_{Ntot} = 860 K). The cause of this phenomenon is due to the size effect. The results show that, as D increases, L increases. T_{Ntot} increases nearly in a linear manner with D according to the approximated formula T_{Ntot} = − 440.6 + 83D (Figure 6g). Through this formula, researchers can adjust the nanoparticle size and subtract the B field to be suitable for specific applications. For example, one can fabricate a nano synthetic thin-film operating at T_{Ntot} = 300 K with the Earth's magnetic field. For this purpose, we study below the influence of magnetic field B, nanoparticle size D on T_{Ntot} at room temperature T = 300 K. The results of the influence of B and D on the Néel phase transition temperature (T_{Ntot}) show that, when increasing both B and D, we have a decrease in Mtot and an increase in Stot. This leads to the conclusion that the magnetization of the material always decreases, and the entropy of materials increases. This is very interesting for future applications of magnetic nanomaterials.

Relationship between B and D at Room Temperature 300 K

Considering the above research results, we investigate the influence at room temperature. We investigate the influence of nanoparticle size at the values of B = 0.025, 0.045, and 0.065 T with dimensions D = 10, 12, and 14 nm (corresponding to L = 114, 174, and 234 nm). The results are shown in Figure 7.

The results show that, when the Fe_2O_3 nano synthetic thin film is placed in the external magnetic field B = 0.025 T with the rotation angle α = 90° for the increase in the size of the nanoparticle D from D = 10 nm (L = 206 nm) to D = 12, 14 nm (L = 244, 282 nm), M_{tot} decreases from M_{tot} = 0.943 to M_{tot} = 0.740, and S_{tot} increases from S_{tot} = −0.597 to S_{tot} = 1.803; whereas for D = 10 nm (L = 206 nm), M_{tot} decreases from M_{tot} = 0.941 to M_{tot} = 0.871, and S_{tot} increases from S_{tot} = −0.875 to S_{tot} = 1.304; with D = 12 nm (L = 244 nm), M_{tot} decreases from M_{tot} = 0.936 to M_{tot} = 0.909, and S_{tot} increases from S_{tot} = −0.908 to S_{tot} = 1.002. For D = 14 nm (L = 282 nm) we also have a decrease in the magnitude of M_{tot} and an increase in S_{tot}. Therefore, as B and D increase, the magnetic phase transition temperature increases from T_{Ntot} = 185 K (Figure 7a1) to T_{Ntot} = 324 K (Figure 7a2), 515 K (Figure 7a3). For B = 0.045 T, when D increases from D = 10 nm (L = 206 nm) to D = 12, 14 nm (L = 244, 282 nm), the magnetic phase transition temperature increases from T_{Ntot} = 220 K (Figure 7b1) to T_{Ntot} = 383 K (Figure 7b2), 596 K (Figure 7b3). With B = 0.065 T, when D increases from D = 10 nm (L = 206 nm) to D = 12, 14 nm (L = 244, 282 nm), the magnetic phase transition temperature increases from T_{Ntot} = 255 K (Figure 7c1) to T_{Ntot} = 445 K (Figure 7c2), and T_{Ntot} > 600 K (Figure 7c3). The obtained results show that, at room temperature, because the magnetic field of the earth, B, is very small, one can increase the nanoparticle size to nearly 12 nm, then T_{Ntot} = T = 300 K. In addition, between T_{Ntot} and D there is a relationship that satisfies the equation T_{Ntot} = −440.6 + 83D. This is a very useful result. In practice, researchers can manufacture Fe$_2$O$_3$ thin films right at ambient conditions (at room temperature). Achieving the size D = 12 nm (L = 244 nm), these thin films can be used not only in magnetic devices.

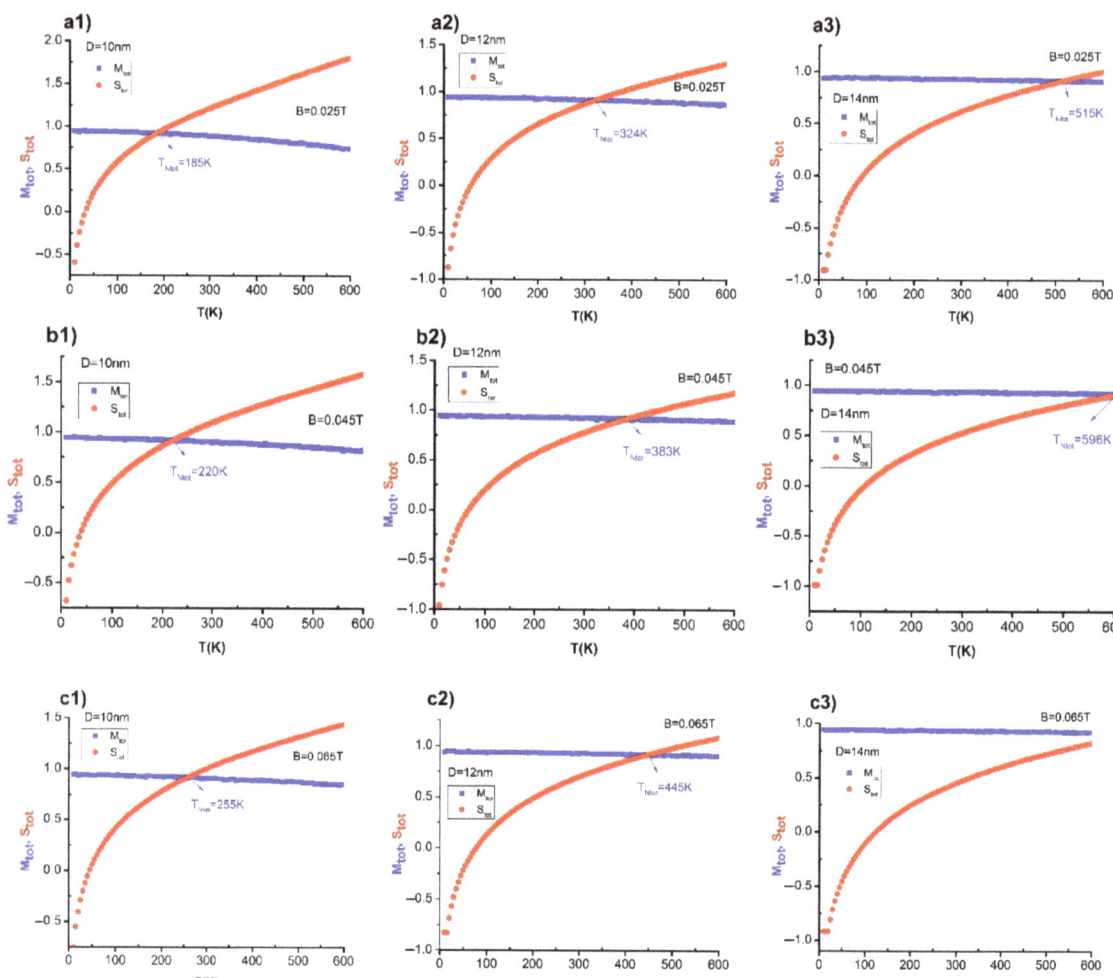

Figure 7. Néel phase transition temperature (T$_{Ntot}$) of Fe$_2$O$_3$ nano synthetic thin film mxm = 20 × 20, in B = 0.1 T with rotation angle α = 90°, for D = 10 nm and different values of B, B = 0.025 T (**a1**), B = 0.045 T (**b1**), B = 0.065 T (**c1**); for D = 12 nm and different values of B, B = 0.025 T (**a2**), B = 0.045 T (**b2**), B = 0.065 T (**c2**); for D = 14 nm and different B, B = 0.025 T (**a3**), B = 0.045 T (**b3**), B = 0.065 T (**c3**).

4. Conclusions

In this study the following results were obtained:
- We successfully studied the influence of the matrix structure (mxm) of thin-film, rotation angle (α), magnetic field (B), and size (D) of Fe$_2$O$_3$ nanoparticle on the magnetic characteristic quantities such as the magnetization-oriented z-direction (M$_{zE}$), z-axis magnetization (M$_z$), total magnetization (M$_{tot}$), and total entropy (S$_{tot}$) of Fe$_2$O$_3$ nanocomposites by Monte-Carlo simulation method.
- We successfully determined the magnetic phase transition temperature Néel (T$_{Ntot}$). The obtained results show that when the mxm increases from mxm = 5 × 5 (L = 27 nm)

to mxm = 10 × 10, 15 × 15, 20 × 20, 30 × 30, 40 × 40 (L = 62, 96, 130, 198, 267 nm), the T_{Ntot} increases slightly from T_{Ntot} = 68 K to T_{Ntot} = 73 K. When the matrix rotation angle α increases from α = 0° to α = 90°, the T_{Ntot} decreases slightly from T_{Ntot} = 93 K to T_{Ntot} = 68 K. The increase in B (from B = 0.1 T to B = 0.9 T) determines an increase in T_{Ntot} (from T_{Ntot} = 68 K to T_{Ntot} = 148, 228, 300, 376 K). The increase in D (from D = 6 nm (L = 130 nm) to D = 8, 10, 12, 14 nm (L = 168, 206, 244, 282 nm)) determines an increase in T_{Ntot} (from T_{Ntot} = 68 K to T_{Ntot} = 164, 320, 560, and higher 600 K). The results show that when B and D increase, T_{Ntot} increases also.
- In addition, between T_{Ntot} and D, there is a linear relationship that satisfies the equation T_{Ntot} = −440.6 + 83D. This is a very interesting result that can be used in practical applications from cooling technology.

Author Contributions: D.N.T.: Conceptualization, Methodology, Investigation, Validation, Resources, Supervision, Writing—original draft—review and editing, Formal analysis. V.C.L.: Validation, Writing and editing. Ș.Ț.: Writing—editing. All authors have read and agreed to the published version of the manuscript.

Funding: This research received no external funding.

Institutional Review Board Statement: Not applicable.

Informed Consent Statement: Not applicable.

Data Availability Statement: The data that support the findings of this study are available from the corresponding author upon reasonable request.

Acknowledgments: We would like to express our deep gratitude to M.R. Dudek for support of our work, especially for giving the simulation program code.

Conflicts of Interest: The funders had no role in the design of the study; in the collection, analyses, or interpretation of data; in the writing of the manuscript, or in the decision to publish the results and the authors declare no conflict of interest.

References

1. Baibich, M.N.; Broto, J.M.; Fert, A.; Dau, F.N.V.; Petroff, F. Giant magnetoresistance of (001)Fe/(001)Cr magnetic superlattices. *Phys. Rev. Lett.* **1988**, *61*, 2472–2475. [CrossRef]
2. Binasch, G.; Grünberg, P.; Saurenbach, F.; Zinn, W. Enhanced magnetoresistance in layered magnetic structures with antiferromagnetic interlayer exchange. *Phys. Rev. B* **1989**, *39*, 4828–4830. [CrossRef] [PubMed]
3. Kosterlitz, J.M.; Thoules, D.J. Ordering, metastability and phase transitions in two-dimensional systems. *J. Phys. C Solid State Phys.* **1973**, *6*, 1181–1203. [CrossRef]
4. Haldane, F.D.M. Nonlinear field theory of large-spin heisenberg antiferromagnets: Semiclassically quantized solitons of the one-dimensional easy-axis neel state. *Phys. Rev. Lett.* **1983**, *50*, 1153–1156. [CrossRef]
5. Dudek, K.K.; Gatt, R.; Dudek, M.R.; Grima, J.N. Controllable hierarchical mechanical metamaterials guided by the hinge design. *Materials* **2021**, *14*, 758. [CrossRef]
6. Dudek, M.R.; Dudek, K.K.; Wolak, W.; Wojciechowski, K.W.; Grima, J.N. Magnetocaloric materials with ultra-small magnetic nanoparticles working at room temperature. *Sci. Rep.* **2019**, *9*, 17607. [CrossRef] [PubMed]
7. Dudek, M.R.; Wojciechowski, K.W.; Grima, J.N.; Caruana-Gauci, R.; Dudek, K.K. Colossal magnetocaloric effect in magneto auxetic systems. *Smart Mater. Struct.* **2015**, *24*, 085027. [CrossRef]
8. Dudek, K.K.; Wolak, W.; Dudek, M.R.; Caruana-Gauci, R.; Gatt, R.; Wojciechowski, K.W.; Grima, J.N. Programmable magnetic domain evolution in magnetic auxetic systems. *Phys. Status Solidi RRL* **2017**, *11*, 1700122. [CrossRef]
9. Dudek, M.R.; Grima, J.N.; Cauchi, R.; Zerafa, C.; Gatt, R.; Zapotoczny, B. Space dependent mean eld approximation modelling. *J. Stat. Phys.* **2014**, *154*, 1508–1515. [CrossRef]
10. Mahhouti, Z.; Ali, M.B.; Moussaoui, H.El.; Hamedoun, M.; Marssi, M.El.; Kenz, A.El.; Benyoussef, A. Structural and magnetic properties of $Co_{0.7}Ni_{0.3}Fe_2O_4$ nanoparticles synthesized by sol-gel method. *Appl. Phys. A* **2016**, *122*, 651–655. [CrossRef]
11. Middey, S.; Jana, S.; Ray, S. Surface spin-glass and exchange bias in Sr_2FeMoO_6 nanoparticle. *J. Appl. Phys.* **2010**, *108*, 043918–043923. [CrossRef]
12. Petracic, O.; Chen, X.; Bedanta, S.; Kleemann, W.; Sahoo, S.; Cardoso, S.; Freitas, P.P. Collective states of interacting ferromagnetic nanoparticles. *J. Magn. Magn. Mater.* **2006**, *300*, 192–197. [CrossRef]
13. Bedanta, S.; Petracic, O.; Kleeman, W. Supermagnetism. In *Handbook of Magnetic Materials*; North Holland: Amsterdam, The Netherlands, 2015; Volume 23, pp. 1–83.

14. Parker, D.; Lisiecki, I.; Pileni, M.P. Do 8 nm Co nanocrystals in long-range-ordered face-centered cubic (fcc) supracrystals show superspin glass behavior? *J. Phys. Chem. Lett.* **2010**, *1*, 1139–1142. [CrossRef]
15. Normile, P.S.; Andersson, M.S.; Mathieu, R.; Lee, S.S.; Singh, G.; De Toro, J.A. Demagnetization efects in dense nanoparticle assemblies. *Appl. Phys. Lett.* **2016**, *109*, 152404. [CrossRef]
16. Seki, T.; Iwama, H.; Shima, T.; Takanashi, K. Size dependence of the magnetization reversal process in microfabricated L10-FePt nano dots. *J. Phys. D Appl. Phys.* **2011**, *44*, 335001. [CrossRef]
17. Bedanta, S.; Seki, T.; Iwama, H.; Shima, T.; Takanashi, K. Superferromagnetism in dipolarly coupled L10 FePt nanodots with perpendicular magnetization. *Appl. Phys. Lett.* **2015**, *107*, 152410. [CrossRef]
18. Veitch, R.J. Soft-magnetic underlayer for MP data tape. *IEEE Trans. Magn.* **2001**, *37*, 1609–1611. [CrossRef]
19. Cornell, R.; Schwertmann, U. (Eds.) Applications. In *The Iron Oxides: Structure, Properties, Reactions, Occurrences and Uses*, 2nd ed.; Wiley-VCH: Weinheim, Germany, 2004; pp. 509–524.
20. Boutahar, A.; Moubah, R.; Lemziouka, H.; Hajjaji, A.; Lassri, H.; Hlil, E.K.; Bessais, L.; Magn, J. Magnetocaloric effect in CoEr$_2$ intermetallic compound. *Magn. Mater.* **2017**, *444*, 106–110. [CrossRef]
21. Mørup, S.; Tronc, E. Superparamagnetic relaxation of weakly interacting particles. *Phys. Rev. Lett.* **1994**, *72*, 3278–3281. [CrossRef] [PubMed]
22. Meisen, U.; Kathrein, H.J.; Imaging, J. The Influence of particle size, shape and particle size distribution on properties of magnetites for the production of toners. *Sci. Technol.* **2000**, *44*, 508–513.
23. Chen, D.H.; Liao, M.H. Preparation and characterization of YADH-bound magnetic nanoparticles. *J. Mol. Catal. B* **2002**, *16*, 283–291. [CrossRef]
24. Dutta, A.K.; Maji, S.K.; Adhikary, B. γ-Fe$_2$O$_3$ nanoparticles: An easily recoverable effective photo-catalyst for the degradation of rose bengal and methylene blue dyes in the waste-water treatment plant. *Mater. Res. Bull.* **2014**, *49*, 28–34. [CrossRef]
25. Takafuji, M.; Ide, S.; Ihara, H.; Xu, Z.H. Preparation of Poly(1-vinylimidazole)-Grafted Magnetic Nanoparticles and Their Application for Removal of Metal Ions. *Chem. Mater.* **2004**, *16*, 1977–1983. [CrossRef]
26. Sourty, E.; Ryan, D.H.; Marchessault, R.H. Characterization of magnetic membranes based on bacterial and man-made cellulose. *Cellulose* **1998**, *5*, 5–17. [CrossRef]
27. Fiorani, D.; Testa, A.M.; Lucari, F.; D'orazio, F.; Romero, H. Magnetic properties of maghemite nanoparticle systems: Surface aniso-tropy and interparticle interaction effects. *Phys. B Condens. Matter* **2002**, *320*, 122–126. [CrossRef]
28. Billotey, C.; Wilhelm, C.; Devaud, M.; Bacri, J.C.; Bittoun, J.; Gazeau, F. Cell internaliza-tion of anionic maghemite nanoparticles: Quantitative effect on magnetic resonance imaging. *Magn. Reson. Med.* **2003**, *49*, 646–654. [CrossRef]
29. Sousa, M.H.; Rubim, J.C.; Sobrinho, P.G.; Tourinho, F.A. Biocompatible magnetic fluid precursors based on aspartic and glu-tamic acid modified maghemite nanostructures. *J. Magn. Magn. Mater.* **2001**, *225*, 67–72. [CrossRef]
30. Lam, U.T.; Mammucari, R.; Suzuki, K.; Foster, N.R. Processing of iron oxide nanoparticles by supercritical fluids. *Ind. Eng. Chem. Res.* **2008**, *47*, 599–614. [CrossRef]
31. Jing, Z.; Wu, S. Synthesis, characterization and gas sen-sing properties of undoped and Co-doped γ-Fe$_2$O$_3$ based gas sensors. *Mater. Lett.* **2006**, *60*, 952–956.
32. Jing, Z.; Wang, Y.; Wu, S. Preparation and gas sensing properties of pure and doped γ-Fe$_2$O$_3$ by an anhy-drous solvent method. *Sens. Actuator B Chem.* **2006**, *113*, 177–181. [CrossRef]
33. Ray, I.; Chakraborty, S.; Chowdhury, A.; Majumdar, S.; Prakash, A.; Pyare, R.; Sen, A. Room tem-perature synthesis of γ-Fe$_2$O$_3$ by sonochemical route and its response towards butane. *Sens. Actuator B Chem.* **2008**, *130*, 882–888. [CrossRef]
34. Gupta, A.K.; Gupta, M. Synthesis and surface engineer-ing of iron oxide nanoparticles for biomedical applications. *Biomaterials* **2005**, *26*, 3995–4021. [CrossRef] [PubMed]
35. Bastow, T.; Trinchi, A.; Hill, M.; Harris, R.; Muster, T. Vacancy ordering in γ-Fe$_2$O$_3$ nanocrystals observed by 57 Fe NMR. *J. Magn. Magn. Mater.* **2009**, *321*, 2677–2681. [CrossRef]
36. Benyettou, F.; Lalatonne, Y.; Sainte-Catherine, O.; Monteil, M.; Motte, L. Superparamagnetic nanovector with anti-cancer prop-erties: γ-Fe$_2$O$_3$@Zoledronate. *Int. J. Pharm.* **2009**, *379*, 324–327. [CrossRef]
37. Forsman, M. Gastric emptying of solids measured by means of magnetised iron oxide powder. *Med. Biol. Eng. Comput.* **1998**, *36*, 2–6. [CrossRef]
38. Yanagihara, H.; Hasegawa, M.; Kita, E.; Wakabayashi, Y.; Sawa, H.; Siratori, K. Iron vacancy ordered γ-Fe$_2$O$_3$ (001) epitaxial films: The crystal struc-ture and electrical resistivity. *J. Phys. Soc. Jpn.* **2006**, *75*, 0547081–0547085. [CrossRef]
39. Wiemann, J.A.; Carpenter, E.E.; Wiggins, J.; Zhou, W.; Tang, J.; Li, S.; Mohan, A. Magnetoresistance of a (γ-Fe$_2$O$_3$)$_{80}$Ag$_{20}$ nanocompo-site prepared in reverse micelles. *J. Appl. Phys.* **2000**, *87*, 7001–7003. [CrossRef]
40. Zboril, R.; Mashlan, M.; Petridis, D. Iron-(III) oxides from thermal processes-synthesis, structural and magnetic properties, Mossbauer spectroscopy characterization, and applications. *Chem. Mater.* **2002**, *14*, 969–982. [CrossRef]
41. Cheng, Z.; Fu, Q.; Duan, H.; Cui, Z.; Xue, Y.; Zhang, W. Size-dependent thermodynamics of structural transition and magnetic properties of nano-Fe$_2$O$_3$. *Ind. Eng. Chem. Res. Ind. Eng. Chem. Res.* **2019**, *58*, 19–8418. [CrossRef]
42. Liu, S.; Zhou, J.; Zhang, L. Effects of crystalline phase and particle size on the properties of plate-like Fe$_2$O$_3$ nanoparticles during γ- to α-phase transformation. *J. Phys. Chem. C* **2011**, *115*, 3602. [CrossRef]
43. Cui, Z.; Duan, H.; Fu, Q.; Xue, Y.; Wang, S. Universal size dependence of integral enthalpy and entropy for solid–solid phase transitions of nanocrystals. *J. Phys. Chem. C* **2017**, *121*, 24831. [CrossRef]

44. Zhang, W.; Xue, Y.; Cui, Z. Effect of size on the structural transition and magnetic properties of nano-CuFe$_2$O$_4$. *Ind. Eng. Chem. Res.* **2017**, *56*, 13760. [CrossRef]
45. Mendili, Y.E.; Bardeau, J.F.; Randrianantoandro, N.; Grasset, F.; Greneche, J.M. Insights into the mechanism related to the phase transition from γ-Fe$_2$O$_3$ to α-Fe$_2$O$_3$ nanoparticles induced by thermal treatment and laser irradiation. *J. Phys. Chem. C* **2012**, *116*, 23785. [CrossRef]
46. Wenger, L.E.; Tsoi, G.M.; Vaishnava, P.P.; Senaratne, U.; Buc, E.C.; Ratna, N.R.; Naik, V.M. Magnetic properties of γ-Fe$_2$O$_3$ nanoparticles precipitated in alginate hydrogels. *IEEE Trans. Magn.* **2008**, *44*, 2760. [CrossRef]
47. Hou, D.L.; Nie, X.F.; Shao, S.X.; Lu, P.; Luo, H. Studies on the magnetic anisotropy and the coercivity of granular γ-Fe$_2$O$_3$ powders. *Phys. Status Solidi* **2015**, *161*, 459. [CrossRef]
48. Wang, J.; Wu, W.; Zhao, F.; Zhao, G.M. Suppression of the N'eel temperature in hydro thermally synthesized α-Fe$_2$O$_3$ nano particles. *J. Appl. Phys.* **2011**, *109*, 056101. [CrossRef]
49. Muñoz, J.L.G.; Romaguera, A.; Fauth, F.; Nogués, J.; Gich, M. Unveiling a new high-temperature ordered magnetic phase in Fe$_2$O$_3$. *Chem. Mater.* **2017**, *29*, 9705–9713. [CrossRef]
50. Namai, A.; Sakurai, S.; Nakajima, M.; Suemoto, T.; Matsumoto, K.; Goto, M.; Sasaki, S.; Ohkoshi, S. Synthesis of an electromagnetic wave absorber for high-speed wireless communication. *J. Am. Chem. Soc.* **2009**, *131*, 1170–1173. [CrossRef] [PubMed]
51. Namai, A.; Yoshikiyo, M.; Yamada, K.; Sakurai, S.; Goto, T.; Yoshida, T.; Ohkoshi, S.I. Hard magnetic ferrite with a gigantic coercivity and high frequency millimetre wave rotation. *Nat. Commun.* **2012**, *3*, 1035. [CrossRef]
52. Ortega, A.L.; Estrader, M.; Alvarez, G.S.; Roca, A.G.; Nogues, J. Applications of exchange coupled bi-magnetic hard/soft and soft/hard magnetic core/shell nanoparticles. *Phys. Rep.* **2015**, *553*, 1–32. [CrossRef]
53. Gich, M.; Frontera, C.; Roig, A.; Fontcuberta, J.; Molins, E.; Bellido, N.; Simon, C.; Fleta, C. Magnetoelectric coupling in ε-Fe$_2$O$_3$ nanoparticles. *Nanotechnology* **2006**, *17*, 687–691. [CrossRef]
54. Gich, M.; Roig, A.; Frontera, C.; Molins, E.; Sort, J.; Popovici, M.; Nogues, J. Large coercivity and low-temperature magnetic reorientation in ε-Fe$_2$O$_3$ nanoparticles. *J. Appl. Phys.* **2005**, *98*, 044307. [CrossRef]
55. Gich, M.; Frontera, C.; Roig, A.; Taboada, E.; Molins, E.; Rechenberg, H.R.; Ardisson, J.D.; Macedo, W.A.A.; Ritter, C.; Hardy, V.; et al. High- and lowtemperature crystal and magnetic structures of ε-Fe$_2$O$_3$ and their correlation to its magnetic properties. *Chem. Mater.* **2006**, *18*, 3889–3897. [CrossRef]
56. Tseng, Y.C.; Souza-Neto, N.M.; Haskel, D.; Gich, M.; Frontera, C.; Roig, A.; van Veenendaal, M.; Nogues, J. Nonzero orbital moment in high coercivity ε-Fe$_2$O$_3$ and low-temperature collapse of the magnetocrystalline anisotropy. *Phys. Rev. B Condens. Matter Mater. Phys.* **2009**, *79*, 094404. [CrossRef]
57. Forestier, H.; Guiot-Guillain, G. A new ferromagnetic variety of iron sesquioxide. *C. R. Hebd. Seances Acad. Sci.* **1934**, *199*, 720–723.
58. Schrader, R.; Buttner, G. A new phase of iron (III)-oxide ε-Fe$_2$O$_3$. *Z. Anorg. Allg. Chem.* **1963**, *320*, 220–234. [CrossRef]
59. Ohkoshi, S.; Namai, A.; Sakurai, S. The origin of ferromagnetism in ε-Fe$_2$O$_3$ and ε-Ga$_x$Fe2-xO3 nanomagnets. *J. Phys. Chem. C* **2009**, *113*, 11235–11238. [CrossRef]
60. Sanchez, J.L.; Serrano, A.; Campo, A.D.; Abuín, M.; Fuente, O.R.; Carmona, N. Sol−gel synthesis and micro-raman characterization of ε-Fe$_2$O$_3$ micro- and nanoparticles. *Chem. Mater.* **2016**, *28*, 511–518. [CrossRef]
61. Dung, N.T. Factors affecting the earth's surface on heterogeneous dynamics of CaSiO$_3$ material. *Mater. Sci. Eng. B* **2020**, *260*, 114648.
62. Tuan, T.Q.; Dung, N.T. Molecular dynamics studies the effects of the earth's surface depth on the heterogeneous kinetics of MgSiO$_3$. *Res. Phys.* **2019**, *15*, 102671.
63. Dung, N.T.; Cuong, N.C.; Van, D.Q. Molecular dynamics studies the effect of structure MgSiO$_3$ bulk on formation process geology of the Earth. *Int. J. Comput. Mater. Sci. Eng.* **2019**, *8*, 1950011. [CrossRef]
64. Warburg, E. Magnetische untersuchungen. *Annu. Phys. Chem.* **1881**, *13*, 141–164. [CrossRef]
65. Pecharsky, V.K.; Gschneidner, K.A., Jr. Giant magnetocaloric effect in Gd$_5$(Si$_2$Ge$_2$). *Phys. Rev. Lett.* **1997**, *78*, 4494–4497. [CrossRef]
66. Wada, H.; Tanabe, Y. Giant magnetocaloric effect of MnAs$_{1-x}$Sb$_x$. *Appl. Phys. Lett.* **2001**, *79*, 3302–3304. [CrossRef]
67. Tegus, O.; Brück, E.; Buschow, K.H.J.; de Boer, F.R. Transition-metal-based magnetic refrigerants for room-temperature applications. *Nature* **2002**, *415*, 150–152. [CrossRef] [PubMed]
68. Fujita, A.; Fujieda, S.; Hasegawa, Y.; Fukamichi, K. Itinerant-electron metamagnetic transition and large magnetocaloric effects in La(Fe$_x$Si$_{1-x}$)$_{13}$ compounds and their hydrides. *Phys. Rev. B* **2003**, *67*, 104416. [CrossRef]
69. Hu, F.X.; Shen, B.G.; Sun, J.R.; Wang, G.J.; Cheng, Z.H. Very large magnetic entropy change near room temperature in LaFe$_{11.2}$Co$_{0.7}$Si$_{1.1}$. *Appl. Phys. Lett.* **2002**, *80*, 826–828. [CrossRef]
70. Balli, M.; Fruchart, D.; Gignoux, D. Optimization of La(Fe, Co)$_{13-x}$Si$_x$ based compounds for magnetic refrigeration. *J. Phys. Condens. Matter* **2007**, *19*, 236230. [CrossRef]
71. Balli, M.; Fruchart, D.; Gignoux, D. The LaFe$_{11.2}$Co$_{0.7}$Si$_{1.1}$C$_x$ carbides for magnetic refrigeration close to room temperature. *Appl. Phys. Lett.* **2008**, *92*, 232505. [CrossRef]
72. Giguère, A.; Földeaki, M.; Gopal, B.R.; Chahine, R.; Bose, T.K.; Frydman, A.; Barclay, J.A. Direct measurement of the "giant" adiabatic temperature change in Gd$_5$Si$_2$Ge$_2$. *Phys. Rev. Lett.* **1999**, *83*, 2262–2265. [CrossRef]
73. Sun, J.R.; Hu, F.X.; Shen, B.G. Comment on "direct measurement of the 'giant' adiabatic temperature change in Gd$_5$Si$_2$Ge$_2$. *Phys. Rev. Lett.* **2000**, *85*, 4191. [CrossRef] [PubMed]
74. Földeaki, M.; Chahine, R.; Bose, T.K.; Frydman, A.; Barclay, J.A. Reply. *Phys. Rev. Lett.* **2000**, *85*, 4192. [PubMed]

75. Balli, M.; Fruchart, D.; Gignoux, D.; Zach, R. The "colossal" magnetocaloric effect in $Mn_{1-x}Fe_xAsMn_{1-x}Fe_xAs$: What are we really measuring? *Appl. Phys. Lett.* **2009**, *95*, 072509. [CrossRef]
76. Liu, G.J.; Sun, J.R.; Shen, J.; Gao, B.; Zhang, H.W.; Hu, F.X.; Shen, B.G. Determination of the entropy changes in the compounds with a first-order magnetic transition. *Appl. Phys. Lett.* **2007**, *90*, 032507. [CrossRef]
77. Muto, T.; Takagi, Y. The theory of order-disorder transitions in alloys. In *Solid State Physics*; Seitz, F., Turnbull, D., Eds.; Springer: Berlin/Heidelberg, Germany, 1955.
78. Fratzl, P.; Penrose, O.; Weinkamer, R.; Zizak, I. Coarsening in the ising model with vacancy dynamics. *Phys. A Stat. Mech. Appl.* **2000**, *279*, 100–109. [CrossRef]
79. Dieter, G.E. *Mechanical Metallurgy*, 3rd ed.; McGraw-Hill Ser. in Materials Science and Engineering; McGraw-Hill: New York, NY, USA; London, UK, 1986; p. 212.
80. Tucek, J.; Zboril, R.; Namai, A.; Ohkoshi, S. ε-Fe_2O_3: An advanced nanomaterial exhibiting giant coercive field, millimeter-wave ferromagnetic resonance, and magnetoelectric coupling. *Chem. Mater.* **2010**, *22*, 6483–6505. [CrossRef]
81. Fuentes-García, J.A.; Diaz-Cano, A.I.; Guillen-Cervantes, A.; Santoyo-Salazar, J. Magnetic domain interactions of Fe_3O_4 nanoparticles embedded in a SiO_2 matrix. *Sci. Rep.* **2018**, *8*, 5096. [CrossRef]
82. Dung, T.N.; Cuong, N.C.; Toan, N.Π.; Hung, P.K. Factors on the magnetic properties of the iron nanoparticles by classical Heisenberg model. *Phys. B* **2018**, *532*, 144–148.
83. Dung, N.T.; Van, C.L. Effects of number of atoms, shell thickness, and temperature on the structure of Fe nanoparticles amorphous by molecular dynamics method. *Adv. Civ. Eng.* **2021**, *9976633*, 1–12.
84. Dung, N.T.; Phuong, N.T. Understanding the heterogeneous kinetics of Al nanoparticles by simulations method. *J. Mol. Struct.* **2020**, *1218*, 128498.
85. Dung, N.T. Z-AXIS deformation method to investigate the influence of system size, structure phase transition on mechanical properties of bulk nickel. *Mater. Chem. Phys.* **2020**, *252*, 123275.
86. Hue, D.T.M.; Coman, G.; Hoc, N.Q.; Dung, N.T. Influence of heating rate, temperature, pressure on the structure, and phase transition of amorphous Ni material: A molecular dynamics study. *Heliyon* **2020**, *6*, e05548.
87. Dung, N.T.; Cuong, N.C.; Van, D.Q.; Tuan, T.Q. Study the effects of factors on the structure and phase transition of bulk Ag by molecular dynamics method. *Int. J. Comput. Mater. Sci. Eng.* **2020**, *09*, 2050016.
88. Dung, N.T.; Van, C.L. Factors affecting the depth of the Earth's surface on the heterogeneous dynamics of $Cu_{1-x}Ni_x$ alloy, x = 0.1, 0.3, 0.5, 0.7, 0.9 by Molecular Dynamics simulation method. *Mater. Today Commun.* **2021**, *102812*, 1–8.
89. Van, C.L.; Van, D.Q.; Dung, N.T. Ab initio calculations on the structural and electronic properties of AgAu alloys. *ACS Omega* **2020**, *5*, 31391–31397.
90. Dung, N.T.; Phuong, N.T. Molecular dynamic study on factors influencing the structure, phase transition and crystallization process of $NiCu_{6912}$ nanoparticle. *Mater. Chem. Phys.* **2020**, *250*, 123075.
91. Dung, N.T.; Cuong, N.C.; Van, D.Q. Study on the effect of doping on lattice constant and electronic structure of bulk AuCu by the density functional theory. *J. Multiscale Model.* **2020**, *11*, 2030001.
92. Hoc, N.Q.; Viet, L.H.; Dung, N.T. On the melting of defective FCC interstitial alloy γ-FeC under pressure up to 100 Gpa. *J. Electron. Mater.* **2020**, *49*, 910–916. [CrossRef]
93. Vu Quoc, T.; Do Ba, D.; Tran Thi Thuy, D.; Nguyen Ngoc, L.; Nguyen Thuy, C.; Vu Thi, H.; Cao Long, V.; Țălu, Ș.; Nguyen Trong, D. DFT study on some polythiophenes containing benzo[d]thiazole and benzo[d]oxazole: Structure and band gap. *Des. Monomers Polym.* **2021**, *24*, 274–284. [CrossRef] [PubMed]
94. Vu, Q.T.; Tran, T.T.D.; Nguyen, T.C.; Nguyen, T.V.; Nguyen, H.; Vinh, P.V.; Nguyen-Tri, P. DFT prediction of factors affecting the structural characteristics, the transition temperature and the electronic density of some new conjugated polymers. *Polymers* **2020**, *12*, 1207. [CrossRef]
95. Evans, K.E.; Nkansah, M.A.; Hutchinson, I.J.; Rogers, S.C. Molecular network design. *Nature* **1991**, *353*, 124. [CrossRef]
96. Landau, D.P.; Binder, K. Introduction. In *A Guide to Monte Carlo Simulations in Statistical Physics*, 4th ed.; Cambridge University Press: Cambridge, UK, 2014; pp. 1–6.
97. Schweika, W. *Disordered Alloys: Diffuse Scattering and Monte Carlo Simulations*; Springer: Berlin/Heidelberg, Germany, 1998; pp. 1–3.
98. Huang, C.; Marian, J. A generalized Ising model for studying alloy evolution under irradiation and its use in kinetic Monte Carlo simulations. *J. Phys. Condens. Matter Inst. Phys. J.* **2016**, *28*, 425201. [CrossRef] [PubMed]
99. Sun, C.Q.; Zhong, W.H.; Li, S.; Tay, B.K. Coordination imperfection suppressed phase stability of ferromagnetic, ferroelectric, and superconductive nanosolids. *J. Phys. Chem. B* **2004**, *108*, 1080–1084. [CrossRef]

Article

Phase Transition and Magnetoelectric Effect in 2D Ferromagnetic Films on a Ferroelectric Substrate

Igor Bychkov [1,*], Sergey Belim [2,*], Ivan Maltsev [1] and Vladimir Shavrov [3]

1. Faculty of Physics, Chelyabinsk State University, 454001 Chelyabinsk, Russia; malts_iv@mail.ru
2. Faculty of Radiophysics, Omsk State Technical University, 644050 Omsk, Russia
3. Kotelnikov IRE RAS, 125009 Moscow, Russia; shavrov@cplire.ru
* Correspondence: bychkov@csu.ru (I.B.); sbelim@mail.ru (S.B.)

Abstract: In this paper, we investigate the behavior of 2D ferromagnetic (FM) films on a ferroelectric (FE) substrate with a periodic structure. The two-dimensional Frenkel–Kontorova (FK) potential simulates the substrate effect on the film. The substrate potential corresponds to a cubic crystal lattice. The Ising model and the Wolf cluster algorithm are used to describe the magnetic behavior of a FM film. The effect of an electric field on a FE substrate leads to its deformation, which is uniform and manifests itself in a period change of the substrate potential. Computer simulation shows that substrate deformations lead to a decrease in the FM film Curie temperature. If the substrate deformations exceed 5%, the film deformations become inhomogeneous. In addition, we derive the dependence of film magnetization on the external electric field.

Keywords: ferromagnetic; magnetoelastic; phase transition; thin films; Frenkel–Kontorova potential

1. Introduction

A wide range of spintronic devices uses the magnetoelectric (ME) effect in two-layer systems. The ME effect allows controlling magnetization with an external electric field. That, in turn, allows managing the conductivity of the system due to the phenomenon of giant magnetoresistance. A giant ME effect occurs in systems consisting of a thin FM film deposited on a FE substrate. One of the ME effect appearance mechanisms in such systems shows up due to the shift in the Curie point in the FM film under the substrate influence. The deformation of the FE substrate occurs in an external electric field. The film is also deformed due to the interaction of the atoms of the film and the substrate. Film deformation due to magnetostrictive phenomena leads to a shift in the Curie temperature. At a constant temperature, the electric field shifts the system position in the phase diagram relative to the phase transition point, which leads to a change in magnetization.

Experiments show that a change in the magnetization of thin films under the substrate influence occurs for various materials. Ultrathin platinum films on a $BaTiO_3$ FE substrate [1] significantly change the value of their magnetic moment upon substrate deformation on an external electric field. The ME effect was observed in Ni films on the same substrate [2]. The use of diffraction and reflection of X-rays in thin Pt/Co/Ta films on various FE substrates [3] in an external electric field allows observing the change in the magnetic domains structure. A large magnitude of the ME effect was noted in thin FM films $CoFe_2O_4/Pb(Zr_{0.52}Ti_{0.48})O_3/LaNiO_3$ on a Pt/Ti/SiO$_2$/Si substrate with the coexistence of FE and FM phases [4]. A study of the change in the magnetic moment in a $Ni_{80}Co_{20}$ film on a Pb(Mg, Nb)O$_3$-PbTiO$_3$ substrate under the influence of an external electric field is represented in [5]. This study revealed that the change in magnetization is a consequence of substrate deformations. A large ME effect of 560 mVcm^{-1}E^{-1} occurs in the Pb(Zr$_{0.52}$Ti$_{0.48}$)O$_3$/LaNiO$_3$/Ni [6] heterostructure, where LaNiO$_3$ serves as a buffer layer. Ba$_{0.9}$Ca$_{0.1}$TiO$_3$/CoFe$_2$O$_4$ thin films on Pt/Ti/SiO$_2$/Si [7] substrates exhibit a fairly large ME effect (82 mVcm^{-1}E^{-1}). FM films La$_{0.7}$Ca$_{0.3}$MnO$_3$ on a FE substrate BaTiO$_3$ demonstrate a Curie temperature decrease both under tension and under compression [8].

Theoretical studies of the ME effect are focused mainly on specific substances. The microscopic model for Al$_2$O$_3$ and MgO substrates [9] became a basis for the investigation of the substrate impact on the magnetic properties of BaCoF$_4$ films. The phase transition temperature and the magnetization of the thin film increase for a compressible substrate, whereas they decrease for a tensile substrate. To study magnetization fluctuations in two-dimensional RMnO$_3$ films (R = Tb, Lu, and Y) on a FE substrate, the modified Landau model was used [10].

Thus, in this paper, we study the FM phase transition and ME effect in two-dimensional films on a FE substrate from a common understanding of the crystal lattice symmetry and the interaction of atoms without reference to a specific substance.

2. Materials and Methods

We explore a two-dimensional FM film and investigate its magnetic properties within the Ising model. The undeformed film has a square lattice with a period of a. The atoms located at the film sites are characterized by the value of the spin S, which can take one of two values ($+1/2$ or $-1/2$). The Hamiltonian of such a system is as follows:

$$H = -\sum J(r_{i,j}) S_i S_j, \qquad (1)$$

where r_{ij} is the distance between atoms i and j. $J(r_{ij})$ is the exchange integral. Generally, the exchange interaction between spins decreases exponentially; thus, one can take into account the summation over the nearest neighbors only. When the film is deformed, the distance between the nearest neighbors changes. Therefore, it is necessary to consider the dependence of the exchange integral on the distance as follows:

$$J(r_{ij}) = J_0 e^{(-|r_{ij}-a|/r_0)}, \qquad (2)$$

where J_0 is the exchange integral in an undeformed FM film. Parameter r_0 determines the rate of the exchange integral decrease with distance and depends on the specific type of substance. This approximation leads to the fact that the effects of long-range action, which are present in some systems, are not taken into account. Long-range effects require considering the interaction not only with the nearest neighbors, but also with the second-nearest ones, and it can lead to a change in the critical temperature. For short-range systems, the exponential law is not universal as well; exchange interactions can be more complex, but this approximation is basic in spin models. Other forms of exchange integrals can be described using additive corrections to the exponential law; still, this is hard to be covered in a single article.

The influence of substrate manifests itself through the interaction between film atoms and substrate surface atoms. The substrate surface atoms affect geometric position of the film atoms, and this influence depends on their arrangement. In our simulations, we model the substrate as a two-dimensional potential with periodically located potential wells. We restrict ourselves to the case of a substrate with a square lattice. We use the two-dimensional FK potential to describe the film interaction with the surface of such a substrate [11].

$$U_{sub} = \frac{A}{2}\left(2 - \cos\left(\frac{2\pi x}{b_x}\right) - \cos\left(\frac{2\pi y}{b_y}\right)\right), \qquad (3)$$

where (x, y) are coordinates of a point on the substrate surface, b_x and b_y are the periods of substrate potential along the OX and OY axis correspondingly, and A is the amplitude of substrate potential. For an undeformed substrate, we assume the square crystal lattice and equal periods along both axes ($b_x = b_y = b$). External electric field leads to the deformation of the substrate crystal along the direction of the strength vector. We assume that deformation depends linearly on the electric field strength E at low values of the tension. Within the computer simulation, we apply an external electric field along the OX axis. Thus, the lattice period changes along the OX axis.

$$b_x = \epsilon b, \tag{4}$$

$$1 - \epsilon = \gamma E, \tag{5}$$

where ϵ is the relative deformation and γ is the ferroelectric constant characteristic of the substance.

In addition to interaction with the substrate, mechanical interaction between atoms occurs in a thin film. As in the FK model, we restrict ourselves to the harmonic approximation. The energy of interaction between atoms has the following form:

$$U_{int} = \frac{g}{2} \sum_{n,m} \left((x_{n+1,m} - x_{n,m} - a)^2 + (y_{n,m+1} - y_{n,m} - a)^2 \right), \tag{6}$$

where g is an elastic constant, and $(x_{n,m}, y_{n,m})$ are the coordinates of the (n, m) atom.

It is necessary to minimize the total energy of the atomic system to determine the equilibrium position of the film atoms.

$$U = U_{int} + U_{sub} \rightarrow min, \tag{7}$$

$$U = \frac{A}{2} \left(2 - \cos\left(\frac{2\pi x_{n,m}}{b_x}\right) - \cos\left(\frac{2\pi y_{n,m}}{b_y}\right) \right) + \frac{g}{2} \sum_{n,m} \left((x_{n+1,m} - x_{n,m} - a)^2 + (y_{n,m+1} - y_{n,m} - a)^2 \right). \tag{8}$$

To search for the equilibrium state of atoms in the film, we use the Monte Carlo method. An unperturbed film with atoms at the sites of a square lattice with a period a is considered the initial state. The method of successive iterations determines the equilibrium state. One iteration consists of each lattice atom performing a test shift by a random vector, the length of which does not exceed $0.1a$. If the new position reduces the total energy of the system, then it is taken. Otherwise, the atom returns to its previous position. In the equilibrium state, the iteration does not change the atomic arrangement.

The arrangement of film atoms on the substrate is determined by the competition between the interaction of film atoms with each other and with the substrate. For an undeformed system, atoms are located at the minima of the substrate potential. The minima are displaced relative to the initial position with a uniform substrate deformation. As a result, for some atoms, the interatomic interaction dominates over the interaction with the substrate, while for others, the substrate plays a more significant role. This leads to different displacements of atoms from the initial state, which manifests itself in the form of inhomogeneous deformations. Inhomogeneous magnetic film deformations with uniform FE substrate deformations along one axis were experimentally observed using X-ray diffraction in [12].

Having determined the atomic positions in the film equilibrium state, we pass to its magnetic properties. We determine magnetization m as the average spin per node.

$$m = \sum_i S_i / N, \tag{9}$$

where N is the total number of atoms in the film.

We use the finite-size scaling theory [13] to determine the temperature of phase transitions. System evolution proceeds according to the Wolf cluster algorithm [14]. We investigate systems of different linear dimensions L. For each of those, the dependence of the fourth-order Binder cumulants on temperature [15] is determined:

$$U_4 = 1 - \frac{\langle m^4 \rangle}{3\langle m^2 \rangle^2}. \tag{10}$$

Angle brackets denote averaging over thermodynamic states. We find the Curie temperature as the intersection of the graphs of the Binder cumulants versus temperature for systems with different linear dimensions.

3. Results

3.1. Computer Simulation

We review the systems with linear dimensions from $L = 18$ to 204 with the help of computer simulation. To satisfy the periodic boundary condition (if $b \neq a$), we must choose different sets of L for each b in such a way that the integer number of the L film atoms has an integer number of potential minima N_p, i.e., $La = N_p b$, where L and N_p integer. The central atom is fixed, and the boundary condition is periodical. The search for the equilibrium state of the film's atoms took 100,000 Monte Carlo steps; the first 100,000 MC steps of the magnetic system evolution are discarded, and the magnetic properties are calculated in the next 50,000 MC steps.

For convenience, we pass to the relative values. We choose the film lattice period as a unit of length $a = 1$. We consider systems with undeformed film in the initial state. This state is realized when the periods of the film and substrate coincide ($b = a$). In this case, the atoms of the film are located at the minima of the substrate potential. An external electric field operates along the OX axis; therefore, substrate deformations also occurred along the OX axis ($b_x = \epsilon, b = \epsilon$). The period of substrate potential along the OY axis is unchanged ($b_y = b = 1$). We choose the unit value for the FE constant ($|\gamma| = 1$). Different values of the FE constant will lead to changes in the electric field strength, while the main dependencies will not change. We perform calculations in the range of E from 0 to 0.1. The compression or substrate stretching depends on the sign of γ. Values $\gamma > 0$ provide substrate compression and $E = 1 - \epsilon$, $\gamma < 0$ lead to substrate stretching and $E = \epsilon - 1$. For the elastic constant, a unit value ($g = 1$) is also chosen. The units for the amplitude of the substrate potential are the same as that for the elastic constant. We perform the calculations for $A = 0.1$, 0.5 and 1.0. Figure 1 shows the dependence of Curie temperature on the external electric field strength.

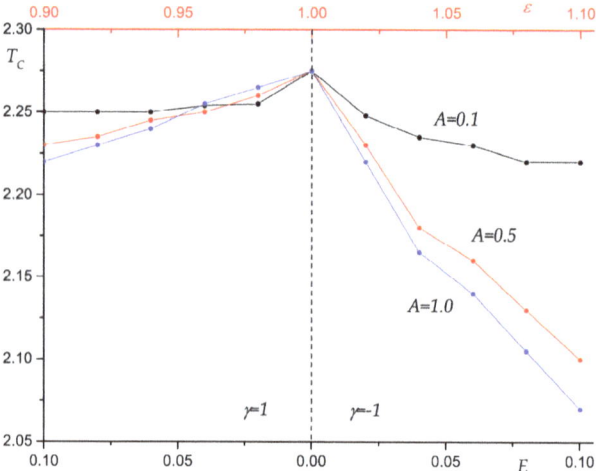

Figure 1. Dependence of the Curie temperature on the strength of the external electric field at different amplitudes of the substrate potential.

It can be seen that substrate deformations under the external electric field lead to a Curie temperature decrease. These results on the Curie temperature behavior are in qualitative agreement with the experimental data [8]. The authors observed a decrease in the phase transition temperature of the epitaxial film $La_{0.7}Ca_{0.3}MnO_3$ upon compression of the ferroelectric substrate BaTiO$_3$ by 1% from 240 K to 160 K. When the substrate was stretched by 6%, the Curie temperature decreased to 195 K. The decrease in the phase transition temperature is associated with rearrangement of the film crystal lattice. Deformations occur unevenly in both cases of compression and tension (Figure 2).

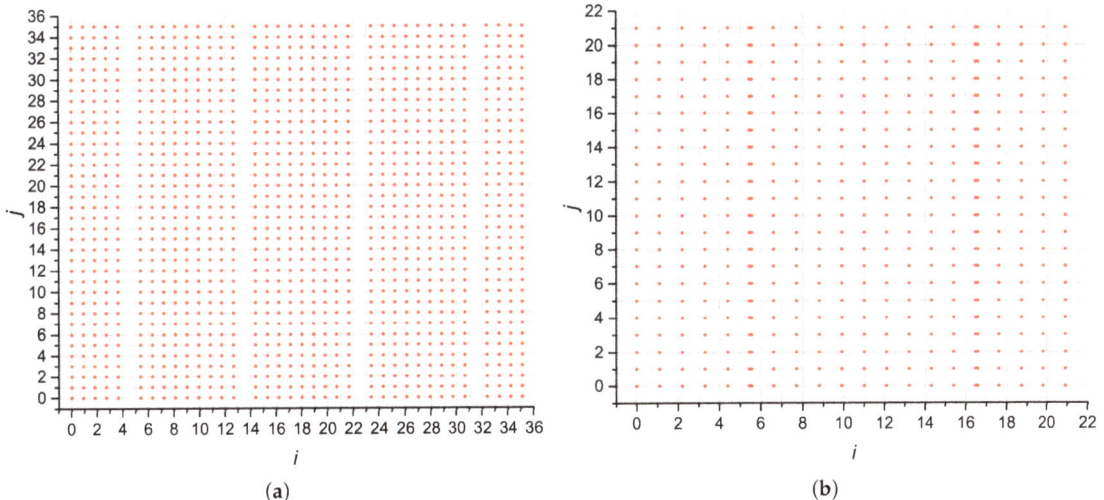

Figure 2. Atomic arrangement of a FM film during tension (**a**) and compression (**b**) of the substrate ($A = 1.0$) (undeforming atomic arrangement falls on the intersection of vertical and horizontal grid lines).

As shown in Figure 2, compression and tension along the OX axis lead to the formation of periodic strip structures in the FM film. In both cases, there is competition between two factors. Upon compression within each stripe, the atomic density increases, which leads to an increase in the Curie temperature. However, the formation of regions of reduced concentration between the stripes leads to a decrease in the phase transition temperature. Stretching inside each stripe decreases the concentration of atoms, leading to a Curie temperature decrease. However, layering of the neighboring stripes leads to an increase in the atoms concentration in the overlap region and an increase in the phase transition temperature. In both cases, the factors lowering the Curie temperature dominate, but the decrease in magnitude is different.

The ME effect consists of a change in the magnetization of the film when the system is in an external electric field. The ME effect is observed in a thin film in the FM phase. Therefore, we consider the system at temperatures below the Curie point. Figure 3 shows the dependence of the thin film magnetization on the strength of the external electric field at three different temperatures.

The magnetization of an FM film decreases in an external electric field, due to the Curie temperature changes. An external electric field brings the system closer to the phase transition point at a fixed temperature, which leads to a decrease in magnetization.

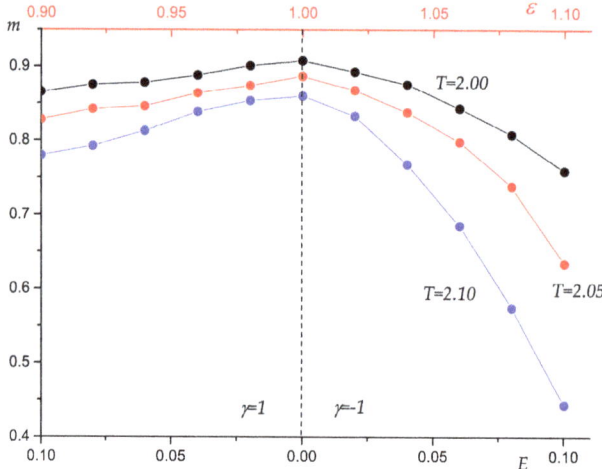

Figure 3. Dependence of the magnetization of a thin film on the strength of the external electric field at three different temperatures for $A = 1.0$.

3.2. Mean-Field Theory

Let us pass to the model of the system under consideration within the mean-field theory. The free energy of the system can be recorded as follows:

$$F = F_M + F_{DF} + F_{MD} + F_{DS} + F_{SF} + F_{FE}. \tag{11}$$

Here, F_M is the free energy of magnetization of a thin FE film, F_{DF} is the deformation energy of a thin film, F_{MD} is the energy of magnetostriction, F_{DS} is the energy of deformation of a FE substrate, F_{SF} is the energy of the film–substrate interaction, and F_{FE} is the energy of interaction of a FE substrate with an external electric field.

Near the second-order phase transition, the magnetization energy term for an FM film can be written as a series in powers of magnetization [16].

$$F_M = AM^2 + BM^4. \tag{12}$$

where B is a positive constant.

$$A = \alpha(T - T_0). \tag{13}$$

where α is a positive constant, T is the system temperature, and T_0 is the intrinsic Curie temperature of the FM film in the absence of external influence.

The free energy with uniform film deformations are as follows:

$$F_{DF} = Cx^2. \tag{14}$$

where x is the change in the linear dimensions of the film, C is the modulus of elasticity of the film.

We can record the energy of elastic deformations of the substrate similarly as follows:

$$F_{DS} = Gy^2. \tag{15}$$

where y is the change in the linear dimensions of the substrate, and G is the elastic modulus of the substrate.

The energy of magnetostriction in most substances is quadratic:

$$F_{MD} = DM^2x. \tag{16}$$

where D is the positive constant of magnetostriction.

The interaction energy of a FE substrate with an external electric field E is linear:

$$F_{FE} = HyE. \tag{17}$$

where H is a positive constant.

At small deformations of the substrate, we can take into account only the linear interaction between the deformations of the substrate and film.

$$F_{SF} = Rxy. \tag{18}$$

where R is a positive constant.

The free energy of a thin FM film on a FE substrate will have the following form:

$$F = AM^2 + BM^4 + Cx^2 + DM^2x + Gy^2 + Rxy + HyE. \tag{19}$$

The condition of the system equilibrium in an external electric field can be expressed as follows:

$$\begin{cases} \frac{\partial F}{\partial M} = 0, \\ \frac{\partial F}{\partial x} = 0, \\ \frac{\partial F}{\partial y} = 0. \end{cases} \tag{20}$$

Hence, we obtain a system of equations for the equilibrium values of deformations (x, y) and magnetization.

$$\begin{cases} 2Gy + HE + Rx = 0, \\ 2Cx + DM^2 + Ry = 0, \\ 2AM + 4BM^3 + 2DMx = 0. \end{cases} \tag{21}$$

Assuming the system is in the FM phase ($M \neq 0$) and solving this system of equations, we obtain the magnetization dependence of a thin film on the electric field strength applied to the substrate.

$$M^2 = \frac{A + DRHE}{D^2 + 2BR^2 - 8BCG}. \tag{22}$$

The general view of magnetization dependence on the electric field strength has the following form:

$$M^2 = \alpha + \beta E. \tag{23}$$

Moreover, $\alpha = \alpha_0(T - T_0)$.

To compare this conclusion with the results of a computer simulation, we plot the dependence of M^2 on E at different temperatures (Figure 4).

The behavior of $M^2(E)$ curves is close to linear at low electric fields. At $E > 0.06$, the dependence ceases to be linear, and the mean-field theory is not applicable. An increase in temperature leads to a parallel displacement of the curves along the OY. The behavior of obtained curves is in qualitative agreement with the conclusions obtained with our model based on the mean-field theory.

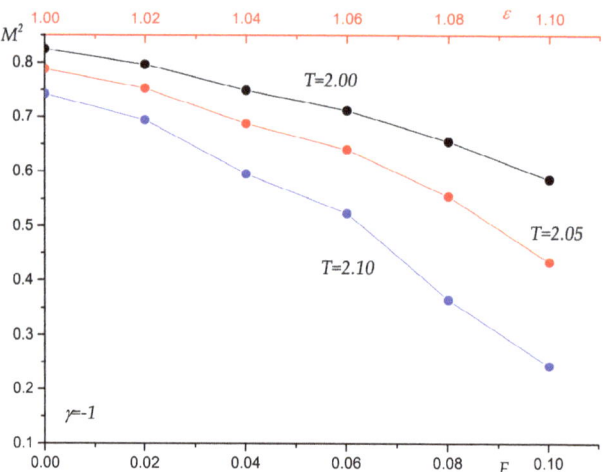

Figure 4. M^2 versus E at different temperatures ($A = 1.0$).

To consider the change in the phase transition temperature in a thin film under the influence of substrate deformations in an external electric field, let us denote the new temperature of the phase transition T_C. The magnetization is zero as an order parameter of the system at the phase transition point. Thus, we can write down the following equation:

$$\alpha(T_C - T_0) + DRHE = 0. \tag{24}$$

The temperature of the phase transition can be recorded as follows:

$$T_C = T_0 - \frac{DRH}{\alpha} E. \tag{25}$$

Thus, an external electric field leads to a Curie temperature decrease in a thin film on a FE substrate. Moreover, the temperature decrease depends linearly on the strength of the electric field. Comparing these results with Figure 1, one can conclude that the mean-field theory is applicable only in weak fields.

4. Discussion

We modeled a computer simulation of the FE substrate influence on the condition and magnetic properties of 2D FM nanofilms. Our results show that the uniform deformations of the substrate lead to inhomogeneous deformations of the film. The interaction between the substrate and film atoms causes film deformations. When the substrate is compressed, it forms strip structures in the film. High-density stripes separate stripes with a low concentration of atoms. Substrate stretching leads to the formation of superstructures with an increased concentration of atoms. Heating or an external electric field can cause substrate deformations. In this case, a substrate-induced structural phase transition takes place.

A change in the film structure leads to changes in its magnetic properties. The Curie temperature is shifting, due to the substrate deformations. It decreases with both substrate compression and stretching. The phase transition temperature changing is caused by the dependence of the exchange integral on the distance between the spins. A decrease in the Curie temperature at any deformation of the substrate is explained by inhomogeneous changes in the atoms arrangement in the film. In both tension and compression, there is a competition of two factors. Areas of increased atomic concentration tend to increase the Curie temperature, while areas of decreased atomic concentration lower it. These trends are associated with the linear dependence of the phase transition temperature on the exchange integral. The regions with low atom concentration are dominant, due to the exponential decay of exchange integral with distance. Long-range effects can significantly change the dependence of Curie temperature on substrate deformations.

The trends in the magnetic characteristics behavior obtained in the simulation agree with the experimental data.

Author Contributions: Conceptualization, V.S.; funding acquisition, I.B. and V.S.; investigation, S.B.; methodology, I.B. and S.B.; project administration, I.B.; software, I.M.; writing—original draft, S.B. and I.M. All authors have read and agreed to the published version of the manuscript.

Funding: This study was funded by Russian Foundation for Basic Researches, project number 20-07-00053, 20-37-70038 and the Ministry of Science and Higher Education of the Russian Federation within the framework of the Russian State Assignment under contract No. 075-00992-21-00. Monte Carlo simulations were performed with the support of the Russian Science Foundation, project number 20-19-00745.

Institutional Review Board Statement: Not applicable.

Informed Consent Statement: Not applicable.

Data Availability Statement: The data presented in this study are available on request from the corresponding author.

Conflicts of Interest: The authors declare no conflict of interest.

References

1. Sun, Q.; Mahfouzi, F.; Velev, J.P.; Tsymbal, E.Y.; Kioussis, N. Ferroelectric-driven tunable magnetism in ultrathin platinum films. *Phys. Rev. Mater.* **2020**, *4*, 124401. [CrossRef]
2. Ghidini, M.; Dhesi, S.S.; Mathur, N.D. Nanoscale magnetoelectric effects revealed by imaging. *J. Magn. Magn. Mater.* **2021**, *520*, 167016. [CrossRef]
3. Chen, A.; Huang, H.; Wen, Y.; Liu, W.; Zhang, S.; Kosel, J.; Sun, W.; Zhao, Y.; Lu, Y.; Zhang, X.-X. Giant magnetoelectric effect in perpendicularly magnetized Pt/Co/Ta ultrathin films on a ferroelectric substrate. *Mater. Horiz.* **2020**, *7*, 2328–2335. [CrossRef]
4. Meenachisundaram, S.; Wakiya, N.; Muthamizhchelvan, C.; Gangopadhyay, P.; Sakamoto, N.; Ponnusamy, S. Enhanced Magnetoelectric Effects in Self-Assembled Hemispherical Close-Packed $CoFe_2O_4$ – $Pb(Zr_{0.52}Ti_{0.48})O_3$ Thin Film. *J. Electron. Mater.* **2021**, *50*, 1699–1706. [CrossRef]
5. Du, W.; Liu, M.; Su, H.; Zhang, H.; Liu, B.; Meng, H.; Xu, G.; Peng, R.; Tang, X. Wide range voltage-impulse-controlled nonvolatile magnetic memory in magnetoelectric heterostructure. *Appl. Phys. Lett.* **2020**, *117*, 222401. [CrossRef]
6. Wang, A.-P.; Song, G.; Zhou, F.-P.; Zhao, L.-N.; Jin, M.; Liu, M.; Zhang, Y.; Hu, L.-L.; Qi, J.; Xu, H.; et al. Strengthened magnetoelectric coupling in $Pb(Zr_{0.52}Ti_{0.48})O_3$/Ni composite through interface modification with $LaNiO_3$ buffer layer. *J. Mater. Sci. Mater. Electron.* **2021**, *32*, 5920–5927. [CrossRef]
7. Shi, M.; Xu, Y.; Zhang, Q.; Yu, Q.; Gu, C.; Zhao, Z.; Guo, L. Impact of heat-treatment conditions on ferroelectric, ferromagnetic and magnetoelectric properties of multi-layered composite films of $Ba_{0.9}Ca_{0.1}TiO_3$/$CoFe_2O_4$. *J. Mater. Sci. Mater. Electron.* **2019**, *30*, 19343–19352. [CrossRef]
8. Ivanov, M.S.; Buryakov, A.M.; Vilarinho, P.M.; Mishina, E.D. Impact of compressive and tensile epitaxial strain on transport and nonlinear optical properties of magnetoelectric $BaTiO_3$-(LaCa)MnO_3 tunnel junction. *J. Phys. D Appl. Phys.* **2021**, *54*, 275302. [CrossRef]
9. Apostolova, I.N.; Apostolov, A.T.; Wesselinowa, J.M. Room temperature ferromagnetism in multiferroic $BaCoF_4$ thin films due to surface, substrate and ion doping effects. *J. Thin Solid Film.* **2021**, *722*, 138567. [CrossRef]
10. Tongue Magne, G.E.; Keumo Tsiaze, R.M.; Fotué, A.J.; Hounkonnou, N.M.; Fai, L.C. Cumulative effects of fluctuations and magnetoelectric coupling in two-dimensional $RMnO_3$ (R = Tb, Lu and Y) multiferroics. *Phys. Lett. A* **2021**, *400*, 127305. [CrossRef]
11. Frenkel, Y.; Kontorova, T. On the theory of plastic deformation and twinning. *Acad. Sci. USSR J. Phys.* **1939**, *1*, 137–149.
12. Nicolenco, A.; Gómez, A.; Chen, X.-Z.; Menéndez, E.; Fornell, J.; Pané, S.; Pellicer, E.; Sort, J. Strain gradient mediated magnetoelectricity in Fe-Ga/P(VDF-TrFE) multiferroic bilayers integrated on silicon. *App. Mater. Today* **2020**, *19*, 100579. [CrossRef]
13. Binder, K. Critical Properties from Monte-Carlo Coarse-Graining and Renormalization. *Phys. Rev. Lett.* **1981**, *47*, 693. [CrossRef]
14. Wolff, U. Collective Monte Carlo Updating for Spin Systems. *Phys. Rev. Lett.* **1989**, *62*, 361. [CrossRef]
15. Landau, D.P.; Binder, K. Phase Diagrams and Multicritical Behavior of a Three-Dimensional Anisotropic Heisenberg Antiferromagnet. *Phys. Rev. B* **1978**, *17*, 2328. [CrossRef]
16. Landau, L. On the Theory of Phase Transitions. *Phys. Z. Sowjet.* **1937**, *11*, 26. [CrossRef]

Effects of Zr/(Zr+Ti) Molar Ratio on the Phase Structure and Hardness of Ti$_x$Zr$_{1-x}$N Films

Jun Zhang [1,2,*], Lijing Peng [1,2], Xiaoyang Wang [1,2], Dongling Liu [1,2] and Nan Wang [1,2]

1 College of Mechanical Engineering, Shenyang University, Shenyang 110044, China; plj1332147938@163.com (L.P.); wxy927@163.com (X.W.); a13804029291@163.com (D.L.); wangnan0602@163.com (N.W.)
2 Key Laboratory of Research and Application of Multiple Hard Films, Shenyang 110044, China
* Correspondence: zhjun88@126.com

Abstract: Ti$_x$Zr$_{1-x}$N hard films with Zr/(Zr+Ti) molar ratios from 20% to 80% were prepared by multi-arc ion plating using any two of elemental Ti, elemental Zr, and TiZr alloy targets. The as-deposited Ti$_x$Zr$_{1-x}$N films displayed similar surface and fracture cross-section morphologies and thicknesses. The effects of Zr/(Zr+Ti) molar ratio on the phase composition, preferred growth orientation, and hardness of the films were discussed. The results showed that the as-deposited films had a face-centered cubic structure and exhibited the typical characteristics of substitutional solid solutions. The lattice constant of the films increased monotonically with increasing Zr/(Zr+Ti) molar ratio. Two preferred growth orientations, corresponding to the two hardness peak values, occurred symmetrically at Zr/Ti molar ratios of 40:60 and 60:40. An inflection point with a small reduction in hardness was observed at the Zr/Ti molar ratio of 50:50.

Keywords: Ti$_x$Zr$_{1-x}$N hard film; Zr/(Zr+Ti) molar ratio; phase structure; preferred growth orientation; hardness

1. Introduction

Hard reactive films of transition element nitrides have been investigated for a long time, and thus significant progress has been made. Several hard multi-component nitride films have been developed. Ti$_x$Zr$_{1-x}$N hard films have attracted attention in recent years because of their high hardness, good red hardness, and simple preparation process. In general, Ti$_x$Zr$_{1-x}$N exhibits higher hardness than TiN and ZrN [1,2]. Zr and Ti atoms can replace each other in the TiN and ZrN lattices to form a TiZrN substitutional solid solution with a face-centered cubic (FCC) structure [3,4]. Ti$_x$Zr$_{1-x}$N films typically display preferred growth along specific crystal plane orientations [1,2,5]. Nevertheless, different preparation methods and deposition parameters lead to inconsistent results in the preferred growth orientations and the peak values of the hardness of Ti$_x$Zr$_{1-x}$N films. For example, the preferred growth of the (111) plane was found in each of the TiN, ZrN, and Ti$_x$Zr$_{1-x}$N films by cathodic arc ion plating [1,6]. Similarly, the preferred growth of the (111) plane was obtained in the Ti$_x$Zr$_{1-x}$N films by reactive sputtering [7,8]. However, the preferred growth orientations in the Ti$_x$Zr$_{1-x}$N films were found to change from the (111) plane (ZrN) to the (200) plane (TiN) as x increased from 0 to 1 when deposited by reactive sputtering [9]. In contrast, the Ti$_x$Zr$_{1-x}$N films deposited by the cathodic arc method were reported to have the opposite trend, where the preferred growth orientations transitioned from the (111) plane (TiN) to the (200) plane (ZrN) as x increased from 0 to 1 [10]. In addition, the effect of bias on preferred growth orientation in Ti$_x$Zr$_{1-x}$N films was reported [11]. A bias that is too low or too high can affect the preferred growth orientation of the films. Similar to the trends for the preferred growth orientation of the Ti$_x$Zr$_{1-x}$N films, the film hardness was found to depend on the composition and the deposition process [1,8,10]. Accordingly, there is no consistent understanding of how the chemical content affects the preferred growth orientation and hardness of $Ti_xZr_{1-x}N$ films owing to the diversity and complexity of deposition processes.

The present work aims to understand the effects of the Zr/Ti molar ratio on the phase structure and hardness of $Ti_xZr_{1-x}N$ films. In particular, the investigation focuses on maintaining constant deposition conditions, microstructure, and thickness of the $Ti_xZr_{1-x}N$ films to achieve consistent results.

2. Materials and Methods

$Ti_xZr_{1-x}N$ films were prepared using a cathodic arc ion plating system (Shenyang Vacuum Technology Research Institute, Shenyang, China) via co-deposition of two cathode targets. Any two of the elemental Ti, elemental Zr, and TiZr (50:50 at.%) alloy targets were co-deposited to vary the Zr/Ti molar ratio in the films. The two cathode targets were set at an angle of 90° relative to the central axis of the coating chamber. The dimensions of the high-speed-steel (W18Cr4V, HRC63-64) substrates used in the deposition were 18 mm × 12 mm × 1.8 mm. After fine polishing, the samples were cleaned by ultrasonication, blow dried, then hung near the central rotating axis of the coating chamber. Each sample was placed symmetrically and equidistantly from the two target surfaces to ensure uniform composition in the films.

The deposition of $Ti_xZr_{1-x}N$ films includes arc bombardment cleaning, alloy transition layer deposition, and $Ti_xZr_{1-x}N$ deposition. Arc bombardment cleaning was carried out by maintaining the two cathode arc currents at 55 A, applying a bias voltage of 350 V, and using an Ar gas flow of 12 sccm for 8 min, when the vacuum reached 8×10^{-3} Pa and the coating chamber temperature reached 220 °C. The deposition of the alloy transition layer was carried out by keeping the two cathode arc currents at 55 A and applying a bias voltage of 180 V while maintaining the Ar gas flow at 12 sccm for 5 min. Finally, $Ti_xZr_{1-x}N$ deposition was carried out by choosing two targets and varying the cathode arc currents, as shown in Table 1, while maintaining a bias voltage of 120 V and N_2 gas flow of 120 sccm for 45 min. The deposition parameters were determined such that a constant temperature of the chamber and a consistent nitrogen content could be maintained during deposition of the $Ti_xZr_{1-x}N$ films.

Table 1. Cathodic arc current combination during the deposition of $Ti_xZr_{1-x}N$ films (total current = 110 A).

Sample No.	Cathodic Arc Current (A)		
	Ti Target	Zr Target	Ti-Zr Alloy Target
1	62	–	48
2	55	–	55
3	48	–	62
4	62	48	–
5	54	56	–
6	48	62	–
7	–	48	62
8	–	62	48

The surface and cross-sectional morphology of the as-deposited $Ti_xZr_{1-x}N$ films were observed using field emission scanning electron microscopy (FE-SEM, S-4800, Hitachi, Japan) equipped with energy-dispersive X-ray spectroscopy (EDS). The surface and cross-sectional compositions were determined using EDS mapping and line scanning, respectively. Phase structure analysis of the as-deposited films was performed using a Rigaku D/max-rA X-ray diffractometer (Dandong Aolong Radiative Instrument Group Co.Ltd, Dandong, China) with CuKα radiation (40 kV and 40 mA). The scanning angle (2θ) ranged from 10 to 80° at 2°/min. The hardness of the as-deposited films was determined using the Vickers micro-indentation test at a load of 0.1 N for 20 s.

3. Results and Discussion
3.1. Surface Morphology and Surface Composition

SEM was adopted to observe the surface morphology of $Ti_xZr_{1-x}N$ hard films. As shown in Figure 1, some "large particles" are randomly distributed on the film layer surface for each of the $Ti_xZr_{1-x}N$ film samples. These particles are generally less than 6 μm in size, and are formed from the micro-droplets sprayed from the cathode target surface during the deposition of the $Ti_xZr_{1-x}N$ films [6,12,13]. Some scattered micro-pits are also observed on the surface of the $Ti_xZr_{1-x}N$ hard films, which are caused by the shedding of the "large particles", as shown in the figure. Similar surface morphologies were observed in all the films, suggesting that the choice of targets had no significant effect on the surface morphology of the films. This finding could be attributed to the constant deposition conditions.

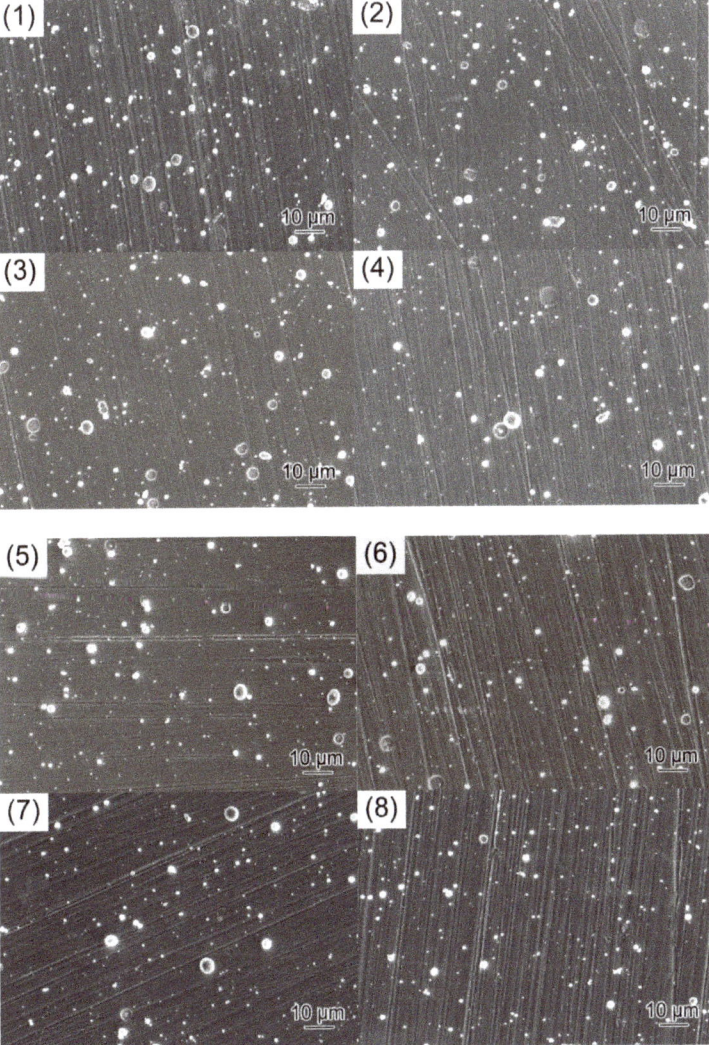

Figure 1. Surface morphologies of the deposited $Ti_xZr_{1-x}N$ films. The numbers (1)–(8) correspond to sample numbers 1–8, respectively.

The surface compositions of the as-deposited $Ti_xZr_{1-x}N$ films under the deposition conditions in Table 1 were characterized using SEM and EDS, and the results are shown in Figure 2. The vertical axis presents the content of each component of the as-deposited film samples. The horizontal axis data is the ratio of metal components, which, from left to right, corresponds to samples 1–8. The nitrogen content in almost all the films was larger than 50 at.% and within 56 at.%–59 at.%, except for sample No. 1. The microstructure and hardness of metal nitride films are affected by the nitrogen to metal component molar ratio in the films [14–17]. In addition, both the phase structure and hardness are very sensitive to the nitrogen content in the films. To ensure the same nitrogen content (at.%) in the films, a relatively high N_2 flow of 120 sccm was selected to supply sufficient N atoms needed to form metal nitride films during the deposition process, despite different arc current combinations forming different targets. As a result, the nitrogen content in the $Ti_xZr_{1-x}N$ hard films was high and basically constant. At the same time, a monotonous change in the Zr/Ti atom ratio was achieved by the control of different cathode target currents.

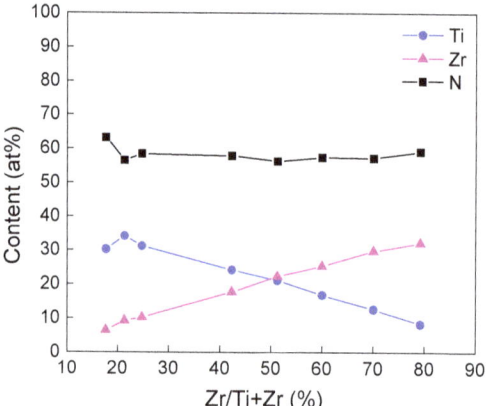

Figure 2. Chemical composition of the surface of the films.

3.2. Cross-Sectional Morphology and Elemental Distribution

The cross-sectional SEM photographs of the as-deposited $Ti_xZr_{1-x}N$ films are shown in Figure 3. The thickness of the films was similar (1 ± 0.1 µm), and all the films exhibited a dense columnar crystal morphology. The consistent thickness and morphology of the films could be attributed to controlled deposition conditions, such as the co-deposition of cathode targets, the small difference between the two arc currents, the sum of the two arc currents kept constant at 110 A, and the constant deposition time of 45 min.

EDS line scanning was employed to observe the elemental distribution in the growth direction of the films, as shown in each inserted figure in Figure 3. The change of relative energy spectrum intensity of Zr and Ti reflects the change of Zr/Ti molar ratio of each film sample, as shown in Figure 3(1)–(8). A gradual increasing trend in the Zr/Ti molar ratio is displayed from sample 1 to sample 8. This is consistent with the surface composition analysis results, as shown in Figure 2. Due to using pure Ar to deposit the alloy transition layer at the initial deposition stage, the energy spectrum intensity of N is zero near the interface between film and substrate and monotonously increases along the film growth direction in each film. The remaining Ar in the coating chamber after the deposition of the transition layer may influence the inflow of N_2 gas. This approximate gradient distribution of nitrogen element could be advantageous in film adhesion [18,19].

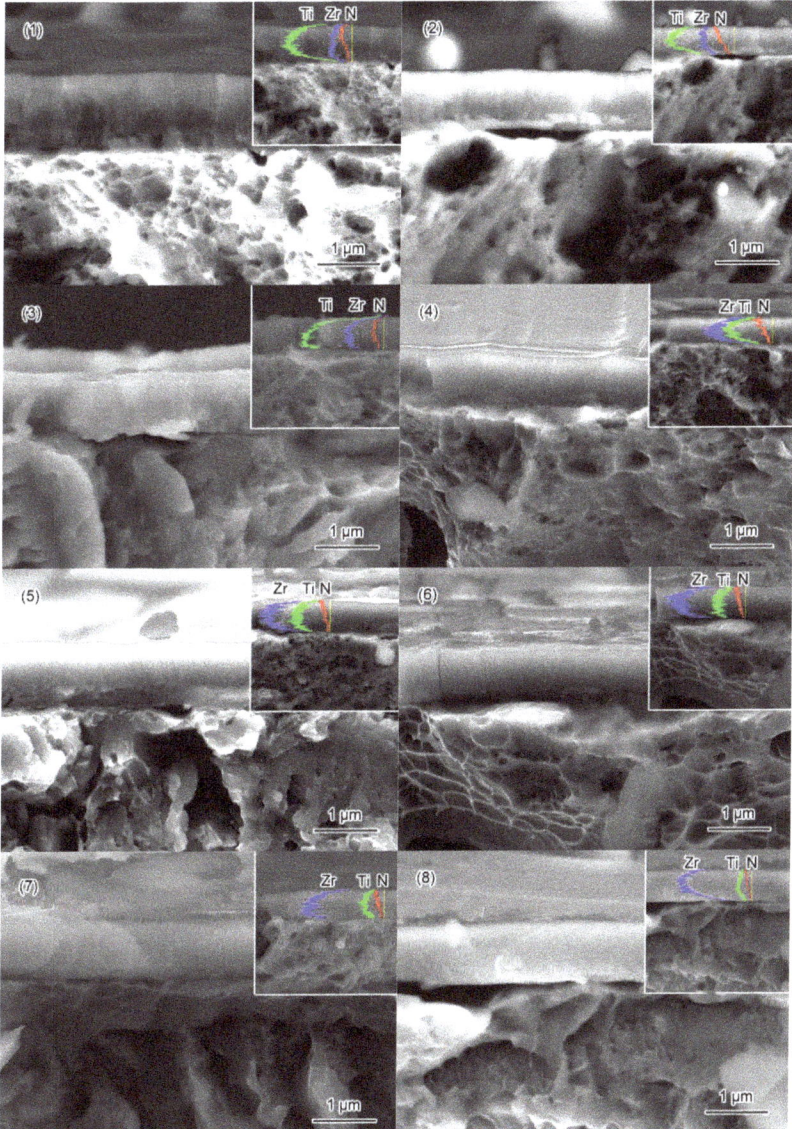

Figure 3. Cross-sectional morphology and elemental distribution in the growth direction of the films. The numbers (1)–(8) correspond to sample numbers 1–8, respectively.

The above discussion shows that the surface morphology, cross-sectional morphology, film thickness, and nitrogen content of the $Ti_xZr_{1-x}N$ hard film samples are consistent with each other, but the relative contents of metal elements (Zr/Ti molar ratio) are different. Therefore, it may be reasonably asserted that in the present study, the relative contents of metal elements (i.e., the Zr/Ti molar ratio) is the only, or the most important, factor influencing the change in the phase composition and mechanical properties of the $Ti_xZr_{1-x}N$ hard films.

3.3. Phase Structure and Preferred Growth Orientation

The XRD patterns of the films are shown in Figure 4. All the films were composed of a nitride phase with an FCC cubic structure and a few metal particles (phase). The 2θ angles of the peaks attributed to the nitride phase are larger than those of the ZrN phase and smaller than those of TiN phase. No ZrN or TiN phase existed in the $Ti_xZr_{1-x}N$ films. Moreover, the diffraction peaks of the films shifted to small angles with an increase in the Zr/Ti molar ratio. Concurrently, the lattice constants of the films increased monotonically. The results suggest that the films were substitutional solid solutions of ZrN and TiN.

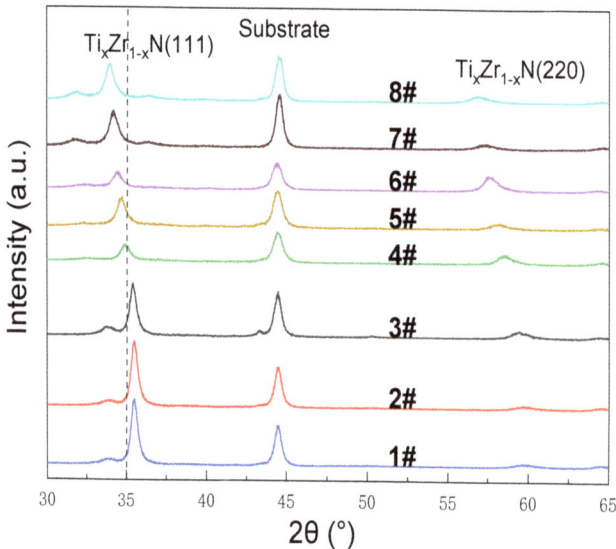

Figure 4. XRD patterns of the films.

Notably, the films exhibited certain textures. The preferred growth orientations of the films varied with an increase in the Zr/Ti molar ratio. When the Zr/Ti molar ratio in $Ti_xZr_{1-x}N$ hard films was lesser than 40:60 or greater than 60:40, the films tended to grow in a preferred orientation on the (111) crystal plane, as shown in Figure 4 (1#, 2#, 3# and 7#, 8#). When the Zr/Ti molar ratio was approximately 40:60 or 60:40, preferred growth orientations in both the (111) and (220) crystal planes were observed, as shown in Figure 4 (4# and 6#). However, when the Zr/Ti molar ratio approached 50:50, the preferred growth orientation returned to the (111) crystal plane. Considering that the properties of the films were similar, we conclude that the Zr/Ti molar ratio is the only, or the most important, factor that influences the preferred growth orientation in the films.

The occurrence of texture in ternary nitride films, such as TiAlN, TiCrN, and AlCrN films, is a common phenomenon. The preferential growth orientations in films of FCC solid solutions vary with the deposition conditions or proportions of metal constituents, or both. Sometimes, the transformation of the preferential growth orientations is accompanied by the occurrence of a second nitride phase; for example, the hexagonal AlN phase in the TiAlN and AlCrN films [20–25]. For $Ti_xZr_{1-x}N$ films, many studies show that the films deposited by ion plating tend to grow in the (111) crystal plane in a wide range of compositions. Examples of such films are the ones with a Zr/Ti molar ratio of 50:50 deposited by cathodic arc ion plating at bias voltages of 120–210 V at 450 °C [3], a molar ratio of 65:35 prepared at 40 V and 400 °C [2], a molar ratio of 66.5:33.5 prepared at 250 V [6], molar ratios from 24:76 to 40:60 prepared at 120 V and at 450–500 °C [4], and molar ratios of 80:20, 70:30, and 40:60 prepared by SCAE technology at 100 V [1]. These results show that $Ti_xZr_{1-x}N$ films prefer to grow in (111) crystal planes because of low substrate bias voltage and low deposition temperature.

However, some inconsistent results regarding the preferred growth orientation of $Ti_xZr_{1-x}N$ films have been reported [9,10]. In one study [10], a series of $Ti_xZr_{1-x}N$ films with Zr/Ti molar ratios of 20:80, 40:60, 50:50, 60:40, and 80:20 was prepared by multi-arc ion plating at a very low bias voltage of 20 V and a high temperature of 600 °C. The research indicates that the preferred growth orientation of the films changed with composition. When the Zr/Zr+Ti molar ratio was very low (<20%), the films displayed the preferred growth orientation in the (111) crystal plane, similar to the TiN film. When the Zr/Zr+Ti molar ratio was between 20% and 40%, the preferred growth orientations were in both the (111) and (200) planes. When the Zr/Zr+Ti molar ratio was further increased, the film's preferred growth orientation was on the (200) plane, similar to the ZrN film. These results indicate that under certain deposition conditions, if the metal constituents of the nitride films have different preferred growth orientations, the preferred growth orientation of the alloy nitride films will change from one preferred growth orientation to the other and experience "double preferred growth orientations" within a certain interval Zr/Ti molar ratio [10]. Another study [9] on $Ti_xZr_{1-x}N$ films deposited by reactive sputtering technology showed that a TiN film had a preferred growth orientation in the (200) crystal plane, while ZrN had a preferred growth orientation in the (111) plane. When the Zr/Zr+Ti molar ratio in the films was between 34% and 45%, the preferred growth orientation was in the (111) and (200) planes. In the other Zr/Zr+Ti molar ratios (between 20% and 34% or between 45% and 67%), the films showed a preferred growth orientation in the (200) plane. Thus, it can be concluded that different deposition techniques and processes significantly influence the preferred growth orientation in $Ti_xZr_{1-x}N$ films. Furthermore, even if the same preparation technique is used, it is challenging to achieve the same deposition temperature and the same thickness and microstructure of the films because of the interaction between the deposition process parameters. For example, both the cathode arc current and the bias voltage influence the deposition temperature. Thus, maintaining the same deposition conditions and the same thickness and microstructure of the films are necessary to investigate the effect of the Zr/Ti molar ratio on the preferred growth orientations and the properties of the $Ti_xZr_{1-x}N$ films.

In general, both ZrN and TiN films prepared by cathode arc ion plating have preferred growth orientation in the (111) crystal planes at conventional bias and deposition temperatures [1,6,26–28]. When the Zr/Ti atom ratio is relatively small or large, the fewer atoms of the substituting metal element do not influence the lattice parameters of the nitride. When the Zr/Ti or Ti/Zr molar ratio approaches 1.1, more defects and a larger lattice distortion are produced due to the difference in atomic radius and electronegativity of the two metals. The preferred growth orientation in the (111) crystal plane observed in ZrN and TiN films is difficult to maintain. A higher index of the crystal plane is selected during growth, and then preferential growth orientations in two crystal planes occur. When the Zr/Ti or Ti/Zr molar ratio is approximately 1:1, the preferred growth orientation of the $Ti_xZr_{1-x}N$ hard film returns to the (111) crystal plane because of short-range ordering in the ZrN and TiN solid solution.

Using Vegard's law, the idealized lattice constant of substitutional solid solutions can be calculated without considering the effect of lattice distortion [29]. The difference between the film's lattice constant measured via XRD and that calculated using Vegard's law can reflect the intensity and density of lattice distortion and lattice defects, respectively. The largest difference in lattice constant occurred when the Zr/Ti molar ratio was approximately 40:60 or 60:40 within the studied composition range, as shown in Figure 5. This result is consistent with the appearance of preferential growth orientations in two crystal planes. When the Zr/Ti molar ratio was approximately 50:50, the difference in the lattice constants decreased slightly. This means that the lattice distortion is less severe, and the strengthening due to the distortion with a Zr/Ti molar ratio of 50:50 may be expected to decrease slightly compared with that of Zr/Ti molar ratios of 40:60 or 60:40.

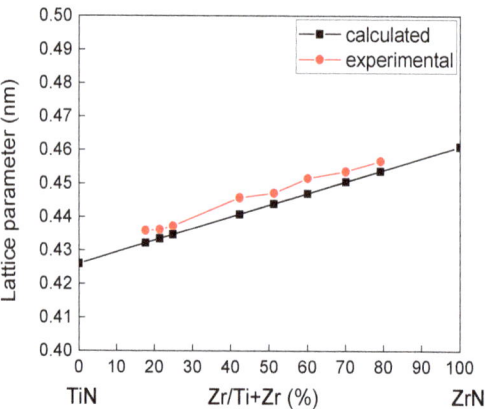

Figure 5. Lattice constant of the films obtained from the experiments and calculations using Vegard's law.

3.4. Hardness

The hardness variation in the films with increasing Zr/Ti molar ratio is shown in Figure 6. When the Zr/Ti or Ti/Zr molar ratio was much lesser than 50:50, the hardness of the films was low and tended to be close to the hardness value of ZrN or TiN. When the Zr/Ti or Ti/Zr molar ratio gradually approached 50:50 and was between 21:79 and 70:30, the hardness of the hard films was relatively high, up to approximately 30 GPa. The maximum values of hardness, 32 and 31 GPa, correspond to the Zr/Ti molar ratios of 40:60 and 60:40, respectively. Notably, when the Zr/Ti molar ratio was approximately 50:50, the hardness of the films decreased slightly, and an inflection point in the hardness value curve was observed. These results can be attributed to the phase structure of the films. The two maximum hardness values correspond to the compositions with Zr/Ti molar ratios of 40:60 and 60:40. These compositions also correspond to the preferred growth orientations in two crystal planes and the difference in the lattice constant obtained from the measurements and Vegard's law.

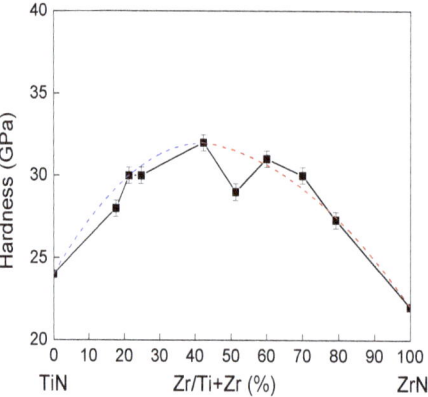

Figure 6. Hardness of the films.

The relationship between the composition and hardness of $Ti_xZr_{1-x}N$ films has been discussed in other related studies. The hardness of films increases monotonically with increasing Zr/Zr+Ti molar ratios from 6:94 to 18:82 to 35:65 [7]; however, the hardness increases significantly with decreasing Zr/Ti molar ratios from 2.9:1 to 1.8:1 [2]. The hardness of the films also increases monotonically with decreasing Zr/Ti molar ratios from

80:20 to 60:40 to 40:60 [1]. In summary, the dependence in hardness values of the films with Zr/Zr+Ti molar ratio is parabolic [10]. A quasi-parabolic trend was also observed in the present study, as shown by the colored dashes in Figure 6, with a maximum hardness value corresponding to the Zr/Ti molar ratio of 40:60 but an inflection point at the Zr/Ti ratio of 50:50.

4. Conclusions

1. The surface and cross-sectional morphologies and the deposition rate of $Ti_xZr_{1-x}N$ films with varying Zr/Ti molar ratios were similar when the deposition conditions such as N_2 gas flow and maintaining the total arc current at 110 A during evaporation of any two of the Ti, Zr, and TiZr targets were maintained.
2. The FCC structure of the films was retained while the lattice constant increased monotonically with increasing Zr/Ti molar ratio, consistent with Vegard's law.
3. The preferred growth orientations of the films were affected by their composition. When the Zr/Ti molar ratio was 40:60 or 60:40, the films showed preferred growth orientation in the (111) and (220) crystal planes; however, at other compositions, the films exhibited a preferred growth only in the (111) crystal plane. A symmetrical distribution in the preferred growth orientation relative to the Zr/Ti molar ratio 50:50 was displayed.
4. The films exhibited a quasi-parabolic hardness distribution. The maximum hardness was observed at a Zr/Ti ratio of 40:60. An inflection point with a small reduction in hardness occurred at a Zr/Ti ratio of 50:50, after which the hardness increased as Zr/Ti ratio increased from 40:60 to 60:40.

Author Contributions: Conceptualization, methodology, J.Z.; investigation, D.L.; resources, X.W.; data curation, writing—original draft preparation, L.P.; writing—review and editing, N.W.; funding acquisition, X.W. All authors have read and agreed to the published version of the manuscript.

Funding: This research was funded by the Liaoning Province Doctor Start-up Fund, Grant No. 2021-BS-276.

Institutional Review Board Statement: Not applicable.

Informed Consent Statement: Not applicable.

Conflicts of Interest: The authors declare no conflict of interest.

References

1. Donohue, L.A.; Cawley, J.; Brooks, J.S. Deposition and characterisation of arc-bond sputter Ti_xZr_yN coatings from pure metallic and segmented targets. *Surf. Coat. Technol.* **1995**, *72*, 128–138. [CrossRef]
2. Grimberg, I.; Zhitomirsky, V.M.; Boxman, R.L.; Goldsmith, S.; Weiss, B.Z. Multicomponent Ti-Zr-N and Ti-Nb-N coatings deposited by vacuum arc. *Surf. Coat. Technol.* **1998**, *108*, 154–159. [CrossRef]
3. Uglov, V.V.; Anishchik, V.M.; Khodasevich, V.V.; Prikhodko, Z.L.; Zlotski, S.V.; Abadias, G.; Dub, S.N. Structural characterization and mechanical properties of Ti-Zr-N coatings, deposited by vacuum arc. *Surf. Coat. Technol.* **2004**, *180*, 519–525. [CrossRef]
4. Uglov, V.V.; Anishchik, V.M.; Zlotski, S.V.; Abadias, G. The phase composition and stress development in ternary Ti-Zr-N coatings grown by vacuum arc with combining of plasma flows. *Surf. Coat. Technol.* **2006**, *200*, 6389–6394. [CrossRef]
5. Rother, B.; Donohue, L.A.; Kappl, H. Quantification of the interface strength between (Ti,Zr)N coatings and high speed steel. *Surf. Coat. Technol.* **1996**, *82*, 214–217. [CrossRef]
6. Du, G.Y.; Ba, D.C.; Tan, Z.; Sun, W.; Liu, K.; Han, Q.K. Vibration damping performance of ZrTiN coating deposited by arc ion plating on TC4 Titanium alloy. *Surf. Coat. Technol.* **2013**, *229*, 172–175. [CrossRef]
7. Jeona, S.; Haa, J.; Choi, Y.; Joc, I.; Leea, H. Interfacial stability and diffusion barrier ability of $Ti_{1-x}Zr_xN$ coatings by pulsed laser thermal shock. *Appl. Surf. Sci.* **2014**, *320*, 602–608. [CrossRef]
8. Huang, J.H.; Chen, Y.F.; Yu, G.P. Evaluation of the fracture toughness of $Ti_{1-x}Zr_xN$ hard coatings: Effect of compositions. *Surf. Coat. Technol.* **2019**, *358*, 487–496. [CrossRef]
9. Sakamoto, I.; Maruno, S.; Jim, P. Preparation and microstructure of reactively sputtered $Ti_{1-x}Zr_xN$ films. *Thin Solid Film.* **1993**, *228*, 169–172. [CrossRef]
10. Hasegawa, H.; Suzuki, T. Effects of second metal contents on microstructure and micro-hardness of ternary nitride films synthesized by cathodic arc method. *Surf. Coat. Technol.* **2004**, *188*, 234–240. [CrossRef]

11. Niu, E.W.; Li, L.; Lv, G.H.; Chen, H.; Li, X.Z.; Yang, X.Z.; Yang, S.Z. Characterization of Ti-Zr-N films deposited by cathodic vacuum arc with different substrate bias. *Appl. Surf. Sci.* **2008**, *254*, 3909–3914. [CrossRef]
12. Harris, S.G.; Doyle, E.D.; Wong, Y.C.; Munroe, P.R.; Cairney, J.M.; Long, J.M. Reducing the macroparticle content of cathodic arc evaporated TiN coatings. *Surf. Coat. Technol.* **2003**, *183*, 283–294. [CrossRef]
13. Wan, X.; Zhao, S.; Yang, Y.; Gong, J.; Sun, C. Effects of nitrogen pressure and pulse bias voltage on the properties of Cr-N coatings deposited by arc ion plating. *Surf. Coat. Technol.* **2009**, *204*, 1800–1810. [CrossRef]
14. Tang, J.F.; Li, C.Y.; Yang, F.C.; Chang, C.L. Influence of nitrogen content and bias voltage on residual stress and the tribological and mechanical properties of CrAlN films. *Coatings* **2020**, *10*, 546. [CrossRef]
15. Lin, Y.W.; Huang, J.H.; Yu, G.P. Effect of nitrogen flow rate on properties of nanostructured TiZrN thin films produced by radio frequency magnetron sputtering. *Thin Solid Film.* **2010**, *518*, 7308–7311. [CrossRef]
16. Tay, B.K.; Shi, X.; Yang, H.S.; Tan, H.; Chua, D.; Teo, S. The effect of deposition conditions on the properties of TiN thin films prepared by filtered cathodic vacuum-arc technique. *Surf. Coat. Technol.* **1999**, *111*, 229–233. [CrossRef]
17. Huang, J.H.; Lau, K.W.; Yu, G.P. Effect of nitrogen flow rate on structure and properties of nanocrystalline TiN thin films produced by unbalanced magnetron sputtering. *Surf. Coat. Technol.* **2005**, *191*, 17–24. [CrossRef]
18. Zhang, J.; Yin, L.Y. Microstructure and mechanical properties of (Ti,Al,Nb)N hard films with N-gradient distributions. *Thin Solid Film.* **2015**, *584*, 141–145. [CrossRef]
19. Chen, L.; Wang, S.Q.; Du, Y.; Li, J. Microstructure and mechanical properties of gradient Ti(C, N) and TiN/Ti(C, N) multilayer PVD coatings. *Mater. Sci. Eng. A* **2007**, *478*, 336–339. [CrossRef]
20. Chang, C.L.; Yang, F.C. Effect of target composition on the microstructural mechanical, and corrosion properties of TiAlN thin films deposited by high-power impulse magnetron sputtering. *Surf. Coat. Technol.* **2018**, *352*, 330–337. [CrossRef]
21. Liu, L.; Zhou, L.; Tang, W.; Ruan, Q.; Li, X.; Wu, Z.; Qasim, A.M.; Cui, S.; Li, T.; Tian, X.; et al. Study of TiAlN coatings deposited by continuous high power magnetron sputtering (C-HPMS). *Surf. Coat. Technol.* **2020**, *402*, 126315–126323. [CrossRef]
22. Liu, J.; Zhu, S.S.; Deng, X.; Liu, J.Y.; Wang, Z.P.; Qu, Z. Cutting performance and wear behavior of AlTiN- and TiAlSiN-coated carbide tools during dry milling of Ti-6Al-4V. *Acta Metall. Sin. (Engl. Lett.)* **2020**, *33*, 459–470. [CrossRef]
23. Sabitzer, C.; Paulitsch, J.; Kolozsvári, S.; Rachbauer, R.; Mayrhofer, P.H. Influence of bias potential and layer arrangement on structure and mechanical properties of arc evaporated Al-Cr-N coatings. *Vacuum* **2014**, *106*, 49–52. [CrossRef]
24. Fan, Q.X.; Zhang, J.J.; Wu, Z.H.; Liu, Y.M.; Zhang, T.; Yan, B.; Wang, T.G. Influence of Al content on the microstructure and properties of the CrAlN coatings deposited by arc ion plating. *Acta Metall. Sin. (Engl. Lett.)* **2017**, *30*, 1221–1230. [CrossRef]
25. Wang, Q.Z.; Zhou, F.; Yan, J.W. Evaluating mechanical properties and crack resistance of CrN, CrTiN, CrAlN and CrTiAlN coatings by nanoindentation and scratch tests. *Surf. Coat. Technol.* **2016**, *285*, 203–213. [CrossRef]
26. Sundgren, J.E. Structure and properties of TiN coatings. *Thin Solid Film.* **1985**, *128*, 21–44. [CrossRef]
27. Molarius, J.M.; Korhonen, A.S.; Ristolainen, E.O. Ti-N phases formed by reactive ion plating. *Vac. Sci. Technol.* **1985**, *3*, 2419–2425. [CrossRef]
28. Larijani, M.; Tabrizi, N.; Norouzian, S.; Jafari, A.; Lahouti, S.; Hosseini, H.H.; Afshari, N. Structural and mechanical properties of ZrN films prepared by ion beam sputtering with varying N_2/Ar ratio and substrate temperature. *Vacuum* **2006**, *81*, 550–555. [CrossRef]
29. Hoerling, A.; Sjölén, J.; Willmann, H.; Larsson, T.; Odén, M.; Hultman, L. Thermal stability, microstructure and mechanical properties of $Ti_{1-x}Zr_xN$ thin films. *Thin Solid Film.* **2008**, *516*, 6421–6431. [CrossRef]

Article

Raman Spectroscopy of V₄O₇ Films

Petr Shvets [1,*], Ksenia Maksimova [1,2] and Alexander Goikhman [1]

[1] Research and Educational Center "Functional Nanomaterials", Baltic Federal University, Aleksandra Nevskogo 14, 236041 Kaliningrad, Russia; xmaksimova@gmail.com (K.M.); agoikhman@kantiana.ru (A.G.)
[2] Deutsches Elektronen-Synchrotron (DESY), Notkestr. 85, 22607 Hamburg, Germany
* Correspondence: pshvets@kantiana.ru

Abstract: A thin film of vanadium oxide Magnéli phase V_4O_7 was produced using cathodic arc sputtering. X-ray diffraction, Rutherford backscattering spectrometry and Raman investigations confirmed the formation of this phase. The Raman spectrum of V_4O_7 differs considerably from the spectrum of another Magnéli oxide, V_3O_5, showing that Raman spectroscopy is an excellent tool for distinguishing between these two phases. Temperature-dependent Raman measurements revealed a significant change of the spectra near the V_4O_7 metal–insulator phase transition.

Keywords: V_4O_7; Magnéli phases; vanadium oxides; thin films; Raman scattering; phase transitions; Rutherford backscatter spectrometry

Citation: Shvets, P.; Maksimova, K.; Goikhman, A. Raman Spectroscopy of V₄O₇ Films. *Coatings* **2022**, *12*, 291. https://doi.org/10.3390/coatings12030291

Academic Editor: Joe Sakai

Received: 23 December 2021
Accepted: 14 February 2022
Published: 22 February 2022

Publisher's Note: MDPI stays neutral with regard to jurisdictional claims in published maps and institutional affiliations.

Copyright: © 2022 by the authors. Licensee MDPI, Basel, Switzerland. This article is an open access article distributed under the terms and conditions of the Creative Commons Attribution (CC BY) license (https://creativecommons.org/licenses/by/4.0/).

1. Introduction

V_4O_7 belongs to a series of vanadium oxides with a general formula of V_nO_{2n-1} (*n* = 3–9) called Magnéli phases [1]. All these phases can be described as modifications of the rutile structure with periodic sheared planes resembling stacking faults into which extra planes of metal atoms are introduced [2]. All Magnéli phases, except V_7O_{13}, undergo a metal–insulator transition with a little discontinuity in lattice constants [3]. For V_4O_7, such transition is observed between 237 and 250 K [3,4]. At room temperature, V_4O_7 is metallic, it has a triclinic structure and may be described in space group $A\bar{1}$ (four vanadium and seven oxygen atoms per unit cell with parameters of a = 5.509 Å, b = 7.008 Å, c = 12.258 Å, α = 95.09°, β = 95.19°, γ = 109.21° at 298 K) [5]. Below 250 K, the resistivity increases discontinuously by a factor of 8–50, the lattice symmetry is unchanged, and there are only slight changes of lattice constants or atomic positions [4,5].

The similarity of structures of Magnéli phases may make it challenging to identify the phase precisely using X-ray diffraction, especially if the sample is nanostructured (for instance, thin film), highly textured and has large lattice stress. Recently, we showed that Raman spectroscopy could be very sensitive to tiny variations of the crystal structure by measuring the spectral features of V_3O_5 around the phase transition at ~420 K [6]. Later, our results were further developed and fully confirmed by the other group [7]. Meanwhile, Raman scattering features for V_4O_7 (as well as for the other Magnéli phases with higher *n*) remain an open question. The only recent report claiming the synthesis and Raman characterization of V_4O_7 demonstrate the typical spectrum of V_2O_5, probably due to sample degradation during the measurement or the presence of impurities, was compiled in [8].

From the practical point of view, V_4O_7 shows potential as a thin film material for smart windows, optical switching and liquid crystal displays [9]. V_4O_7 nanoflakes demonstrate good performance as a photocatalyst [8], V_4O_7 nanowires may be used as optical sensors [10], and V_4O_7 nanocrosses are considered a promising candidate as a cathode for lithium-ion batteries [11].

Typically, V_4O_7 is produced from a mixture of V_2O_3 and V_2O_5 powders by vacuum heating [1,3] or vapor transport using $TeCl_4$ [4,5,12]. Alternatively, this Magnéli phase can be obtained by solvothermal methods [8–11]. Direct sputtering of vanadium in an oxygen atmosphere can also yield V_4O_7, but such reports are scarce [13]. In our work, we used cathodic arc sputtering of vanadium target in reactive oxygen, and under certain conditions, V_4O_7 films were formed. We studied these films using Raman spectroscopy at room and low temperatures and showed that the phase transition significantly influenced the observed spectra.

2. Materials and Methods

Vanadium oxide films were deposited by a cathodic arc sputtering of a vanadium target in an oxygen atmosphere, similarly to our previous work [6]. The substrate was 10 mm × 10 mm (111)-oriented Si crystal with ~200 nm thermally grown oxide layer SiO_2. The substrate temperature during the deposition was 600 °C, and the pressure of reactive oxygen was 0.045 Pa. The arc current was 55 A. The sample was placed about 50 cm away from the cathode, at the point where growth speed was about 17 nm/min. The total deposition time was 15 min, yielding a thickness of about 250 nm.

An X-ray diffraction (XRD) study of the films in θ–2θ geometry was performed using a Bruker AXS D8 DISCOVER setup with a Cu Kα (0.15418 nm) radiation (Bruker, Karlsruhe, Germany). Temperature-dependent measurements were performed using the Anton Paar temperature control unit in the range of −120~+100 °C.

Raman scattering measurements were performed using a Horiba Jobin Yvon micro-Raman spectrometer LabRam HR800 with an ×100 magnification objective lens (numerical aperture of 0.9, Horiba, Villeneuve d'Ascq, France). Measurements were conducted at room and lower temperatures in the argon environment to avoid water condensation on the sample. Peltier element supplied by Kryotherm, Saint-Petersburg, Russia, was used to achieve temperatures of up to −75 °C. A He-Ne laser with a 632.8 nm wavelength was used to excite Raman scattering. Laser power on the sample was 0.5 mW, and a spot diameter was about 10 μm (it was intentionally defocused to avoid sample degradation and reduce overheating). The total acquisition time for each sample was at least 1 h.

Stoichiometry of produced samples was investigated by Rutherford backscattering spectrometry (RBS) using the Van de Graaff accelerator AN2500 (High Voltage Engineering Europa B.V., Amersfoort, The Netherlands) located at Immanuel Kant Baltic Federal University (Kaliningrad, Russia). A sample was exposed to a 1.5 MeV $4He^+$ beam. An angle between the beam and normal to the sample surface was 5°, an exit angle was 21°, and a scattering angle was 160°. A beam current and a total accumulated charge were about 20 nA and 65 μC, respectively. Experimental spectra were processed using SIMNRA software (version: 7.03) [14].

Morphology of the samples was investigated using the atomic force microscope (AFM) SmartSPM Aist-NT in the tapping mode (amplitude 25 nm, AIST-NT Inc., Novato, CA, USA) and the scanning electron microscope (SEM) Zeiss Cross Beam XB 540 (Carl Zeiss Microscopy GmbH, Oberkochen, Germany), which is part of a unique scientific facility «SynchrotronLike», operated at 3 kV.

3. Results and Discussion

XRD pattern for a powder V_4O_7 sample was calculated based on the crystal structure from ref. [5] using Profex software [15] (Figure 1a). Our V_4O_7 film (Figure 1b) is highly textured, and the only strong peak is located at 2θ = 39.04°. The nearest reflection for the bulk sample is $(2\bar{2}\bar{2})$ at 2θ = 38.86°, so the compressive stress in the film is about 0.4%. A $(2\bar{2}\bar{2})$ reflection may be rewritten in rutile VO_2 (R) subcell coordinates as $(200)_R$ [5]. Such texturing is typical for VO_2 films [16,17], including samples produced by arc sputtering [18].

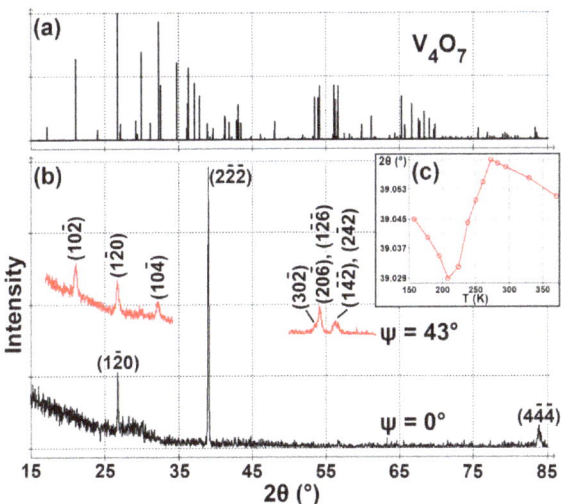

Figure 1. (a) Calculated XRD powder pattern of V_4O_7. (b) XRD patterns of a V_4O_7 film measured at different sample tilts (ψ). (c) Dependence of $(2\bar{2}2)$ peak position on temperature.

One of the characteristic reflections of V_4O_7 is $(10\bar{2})$ at $2\theta = 21.14°$ (for V_5O_9 the equivalent reflection is $(10\bar{2})$ at $2\theta = 22.36°$ [19] and for V_3O_5 the equivalent reflection is (200) at $2\theta = 19.13°$ [20]). For a sample with a texture of $(2\bar{2}2)$, this reflection can be observed in θ–2θ scans if a sample tilt of $\psi = 43°$ is applied (this angle is calculated according to ref. [21]). In Figure 1b we indeed see such a reflection at $2\theta = 21.13°$, confirming the formation of V_4O_7. The experimental configuration at $\psi = 43°$ supports the observation of several other expected reflections: $(1\bar{2}0)$ at $2\theta = 26.81°$ (expected at $2\theta = 26.76°$, $\psi = 32°$), $(10\bar{4})$ at $2\theta = 32.22°$ (expected at $2\theta = 32.3°$, $\psi = 49°$), $(30\bar{2})$ at $2\theta = 53.6°$ (expected at $2\theta = 53.45°$, $\psi = 45°$), $(20\bar{6})$ or $(1\bar{2}6)$ at $2\theta = 54.2°$ (expected at $2\theta = 53.96°$, $\psi = 45°$ or $2\theta = 54.16°$, $\psi = 43°$, respectively), $(1\bar{4}2)$ at $2\theta = 56.25°$ (expected at $2\theta = 56.13°$, $\psi = 43°$) and $(2\bar{4}2)$ at $2\theta = 56.7°$ (expected at $2\theta = 56.63°$, $\psi = 43°$). Since we observe all the expected peaks (with relatively high intensities) and only the expected reflections, we can conclude that our film is indeed V_4O_7 and contains no other phases.

Temperature dependence of the position of $(2\bar{2}2)$ reflection is shown in Figure 1c. We can see the abnormal expansion of the lattice during the sample cooling from 0 to $-65\,°C$. Such behavior is associated with the phase transition in V_4O_7. However, the discontinuity on the graph, which could help to locate the exact transition temperature, can hardly be seen even for single crystals [5]. For thin films, the phase transition is usually not abrupt, which is related to the complex internal strain dynamics in the film [7,22]. Thus, we can only conclude that the phase transition temperature in our sample is above $-65\,°C$.

The formation of V_4O_7 is further confirmed by elemental analysis performed by RBS (Figure 2). An experimental spectrum was fitted, suggesting structures with different values of x in the VO_x layer. The number of incident particles was adjusted to minimize the quadratic deviation for the signal from vanadium (channels 245–280). Then, the value of the reduced quadratic deviation χ_R^2 was analyzed for the oxygen region (channels 75–115). The film composition of V_4O_7 gives the best agreement between experiment and simulation ($\chi_R^2 = 1.4$). The accuracy of the method rules out the formation of V_3O_5 or Magnéli phases with $n \geq 6$ ($\chi_R^2 = 8.6$ and 5.3 for V_3O_5 and V_6O_{11}, respectively). However, it would be challenging to distinguish between V_4O_7 and V_5O_9, since V_5O_9 also gives an acceptable agreement between experiment and simulation ($\chi_R^2 = 2.3$).

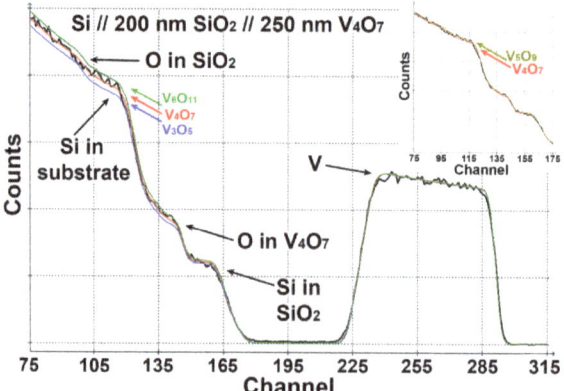

Figure 2. RBS of V_4O_7 film. Three color lines correspond to fits for films with different stoichiometry. Inset shows the comparison between experiment and simulation for V_4O_7 and V_5O_9. 1 channel = 3.94 keV.

Morphology of V_4O_7 film was investigated using SEM and AFM (Figure 3). From the Scherrer equation (FWHM of $(2\bar{2}2)$ peak is 0.15°), we can estimate the average crystallite size as 60 nm. This value is consistent with visible grain sizes (Figure 3a,b,d). Additionally, using a watershed algorithm, we determined that the distribution of equivalent disk diameter for grains has a maximum of 60 nm and a mean value of 100 nm. From the cross-section image (Figure 3c), we can estimate the film thickness as 240 nm, which is in good agreement with the value obtained by RBS.

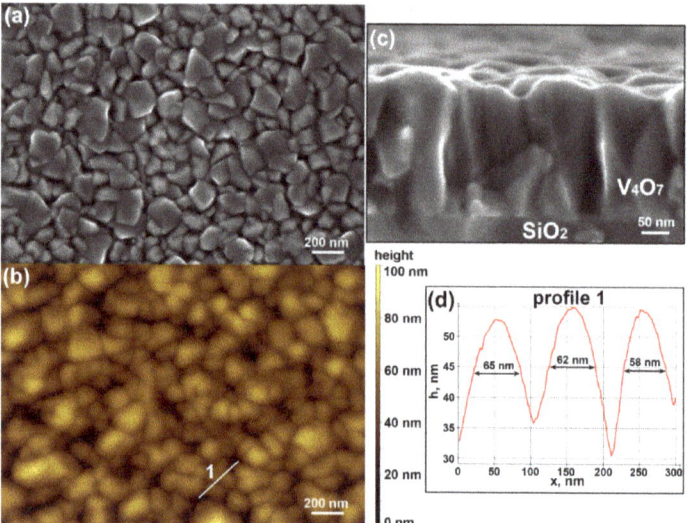

Figure 3. Morphology of V_4O_7 film. (**a**) SEM and (**b**) AFM images of the sample surface. (**c**) SEM image of the sample cross-section. (**d**) Height profile measured by AFM along the white line 1 in (**b**).

Raman spectrum of V_4O_7 film is shown in Figure 4. The spectrum is similar to one observed in our previous work (VOx, Figure 12 in ref. [6]). However, this VO_x phase was not properly identified, and the sample contained impurities. V_4O_7 undergoes a phase transition between −36 and −23 °C, so two different spectra are presented: above (20 °C, Figure 4b) and below (−75 °C, Figure 4d) the phase transition temperature. Cooling the V_4O_7 sample leads to a blueshift of all Raman bands. Moreover, low-temperature modification has four additional peaks at 166, 556, 596 and 714 cm^{-1}. Such spectral changes can be compared with the behavior of another Magnéli phase, V_3O_5 (Figure 4a,c). Similarly to V_4O_7, heating of V_3O_5 above the phase transition leads to the disappearance of strong peaks at higher wavenumbers: 564, 591 and 740 cm^{-1}.

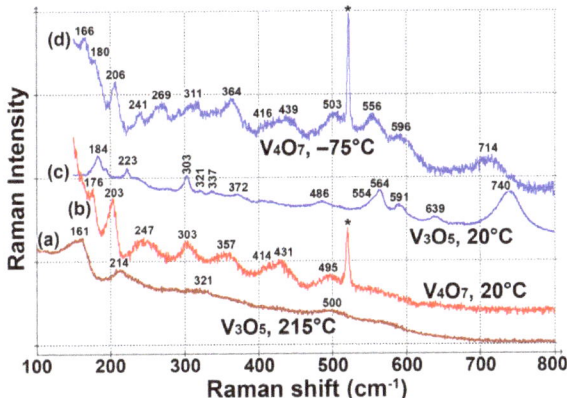

Figure 4. Raman spectra of vanadium oxide Magnéli phases: (**a**) V_3O_5, high-temperature modification, (**b**) V_4O_7, high-temperature modification, (**c**) V_3O_5, low-temperature modification, (**d**) V_4O_7, low-temperature modification. The asterisk marks the position of the Si substrate peak.

Temperature-dependent Raman measurements are summarized in Figure 5. Only slight blueshifts of the bands are observed up to −50 °C. At −55 °C, we can see the appearance of new peaks at 165, 553, 593 and 709 cm^{-1}. Further cooling does not significantly influence the spectrum. The spectral changes are fully reversible if the sample is heated to room temperature after the cooling. From our measurements, we can conclude that the sample undergoes a phase transition at a temperature between −50 and −55 °C. This temperature is slightly lower than the phase transition temperature of bulk V_4O_7 (the reported values are −36 [4] or −23 °C [5]). The discrepancy can be explained by the effect of external surface and grain-boundary surface energies [23]. This effect shifts the phase transition temperature by up to several tens of degrees to lower or higher values and is typical for thin films. Note that the phase transition in other Magnéli phases occurs at much lower temperatures (the highest one is about −100 °C for V_6O_{11} [3]).

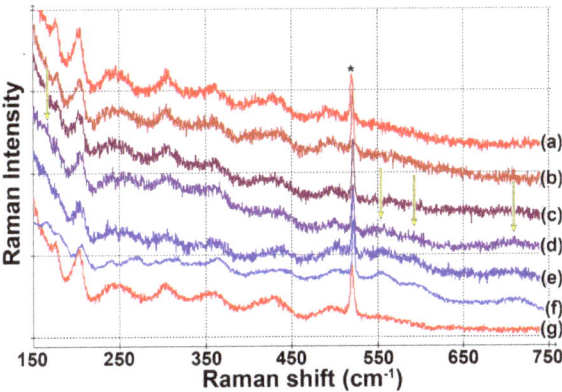

Figure 5. Raman spectra of V_4O_7 at different temperatures: (**a**) 20 °C, (**b**) −40 °C, (**c**) −50 °C, (**d**) −55 °C, (**e**) −60 °C, (**f**) −75 °C, (**g**) 20 °C after low-temperature measurements. The asterisk marks the position of the Si substrate peak. Arrows mark the appearance of new peaks indicating the phase transition.

Temperature dependence of the V_4O_7 film electrical conductivity is given in Figure 6. Similar to XRD, we can see a broad transition region between −25 and −55 °C. The absolute value of conductivity at room temperature agrees well with the value previously reported for the bulk crystal [4]. For the low-temperature region, our curve is shifted by 20–25 K. Such shift is consistent with our observations derived from Raman spectroscopy.

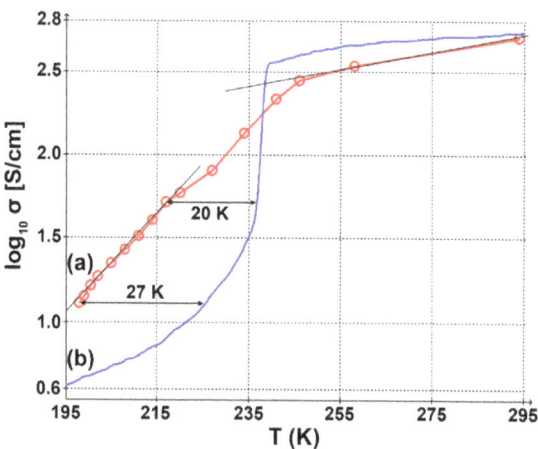

Figure 6. Temperature dependence of electrical conductivity (logarithmic scale). (**a**) V_4O_7 film, (**b**) V_4O_7 single crystal (data from [4]).

4. Conclusions

Magnéli phases of vanadium oxides were directly produced by a sputtering technique. During the cathodic arc sputtering of the vanadium target, V_4O_7 grows by 17 nm/min at 600 °C and 0.045 Pa of oxygen pressure. Produced V_4O_7 films have a preferred orientation of ($2\overline{2}2$) corresponding to (200) in rutile coordinates. The stoichiometry of V_4O_7 was confirmed by Rutherford backscatter spectrometry. The room temperature Raman spectrum of this phase is relatively weak, and its measurement requires careful tuning of the laser power on the sample. The spectrum consists of one strong narrow peak at 203 cm^{-1} and several weaker and broader bands at 176, 247, 303, 357, 414, 431 and 495 cm^{-1}. Metal–insulator phase transition in V_4O_7 was observed by Raman spectroscopy between −50 and −55 °C. The low-temperature spectrum is characterized by four additional bands at 166, 556, 596 and 714 cm^{-1} (at −75 °C).

Author Contributions: Conceptualization, P.S. and K.M.; Data curation, P.S.; Formal analysis, P.S.; Funding acquisition, A.G.; Investigation, P.S.; Methodology, P.S.; Project administration, A.G.; Resources, P.S.; Software, P.S.; Supervision, K.M.; Validation, P.S., K.M. and A.G.; Visualization, P.S.; Writing—Original Draft, P.S.; Writing—Review and Editing, K.M. and A.G. All authors have read and agreed to the published version of the manuscript.

Funding: This research was funded by Ministry of Science and Higher Education of the Russian Federation, Grant No. FZWM-2020-0008.

Institutional Review Board Statement: Not applicable.

Informed Consent Statement: Not applicable.

Data Availability Statement: Data are contained within the article.

Acknowledgments: We are grateful to Ivan Lyatun for performing SEM measurements. We are also grateful to Olga Dikaya for the help with AFM data processing.

Conflicts of Interest: The authors declare no conflict of interest. The funders had no role in the design of the study; in the collection, analyses, or interpretation of data; in the writing of the manuscript, or in the decision to publish the results.

References

1. Andersson, G. Studies on vanadium oxides. I. phase analysis. *Acta Chem. Scand.* **1954**, *8*, 1599–1606. [CrossRef]
2. Horiuchi, H.; Tokonami, M.; Morimoto, N.; Nagasawa, K.; Bando, Y.; Takada, T. Crystallography of V_nO_{2n-1} ($3 \leq n \leq 8$). *Mater. Res. Bull.* **1971**, *6*, 833–843. [CrossRef]
3. Kachi, S.; Kosuge, K.; Okinaka, H. Metal-insulator transition in V_nO_{2n-1}. *J. Solid State Chem.* **1973**, *6*, 258–270. [CrossRef]
4. Andreev, V.N.; Klimov, V.A. Specific features of the electrical conductivity of V_4O_7 single crystals. *Phys. Solid State* **2009**, *51*, 2235. [CrossRef]
5. Marezio, M.; McWhan, D.B.; Dernier, P.D.; Remeika, J.P. Structural aspects of the metal-insulator transition in V_4O_7. *J. Solid State Chem.* **1973**, *6*, 419–429. [CrossRef]
6. Shvets, P.; Dikaya, O.; Maksimova, K.; Goikhman, A. A review of Raman spectroscopy of vanadium oxides. *J. Raman Spectrosc.* **2019**, *50*, 1226–1244. [CrossRef]
7. Lysenko, S.; Rúa, A.; Kumar, N.; Lu, J.; Yan, J.-A.; Theran, L.; Echeverria, K.; Ramos, L.; Goenaga, G.; Hernández-Rivera, S.P.; et al. Raman spectra and elastic light scattering dynamics of V_3O_5 across insulator–metal transition. *J. Appl. Phys.* **2021**, *129*, 025111. [CrossRef]
8. Al-Alharbi, L.; Alrooqi, A.; Ibrahim, M.M.; El-Mehasseb, I.M.; Kumeria, T.; Gobouri, A.; Altalhi, T.; El-Sheshtawy, H.S. In situ H2O2 generation for tuning reactivity of V_4O_7 nanoflakes and V_2O_5 nanorods for oxidase enzyme mimic activity and removal of organic pollutants. *J. Environ. Chem. Eng.* **2021**, *9*, 105044. [CrossRef]
9. Nasr, M.; Gomaa, H.M.; Yahia, I.S.; Saleh, H.A. Novel thermochromic (TC) and electrochromic (EC) characteristics of the V_4O_7 liquid crystal for LCDs and versatile optoelectronic applications. *J. Mol. Liq.* **2021**, *330*, 115620. [CrossRef]
10. Xu, J.; Hu, C.; Han, H.; He, M.; Wan, B.; Xia, C. The synthesis and photoelectric response of single-crystalline V_4O_7 nanowires. In Proceedings of the 2010 3rd International Nanoelectronics Conference, Hong Kong, China, 3–8 January 2010; pp. 413–414. [CrossRef]
11. Wang, X.; Zheng, S.; Mu, X.; Zhang, Y.; Du, H. Additive-free synthesis of V_4O_7 hierarchical structures as high performance cathodes for lithium ion batteries. *Chem. Commun.* **2014**, *50*, 6775–6778. [CrossRef]
12. Demeter, M.; Neumann, M.; Reichelt, W. Mixed-valence vanadium oxides studied by XPS. *Surf. Sci.* **2000**, *454–456*, 41–44. [CrossRef]
13. Razavi, A.; Hughes, T.; Antinovitch, J.; Hoffman, J. Temperature effects on structure and optical properties of radio-frequency sputtered VO_2. *J. Vac. Sci. Technol. A* **1989**, *7*, 1310–1313. [CrossRef]
14. Mayer, M. SIMNRA, a simulation program for the analysis of NRA, RBS and ERDA. *AIP Conf. Proc.* **1999**, *475*, 541–544. [CrossRef]
15. Doebelin, N.; Kleeberg, R. Profex: A graphical user interface for the Rietveld refinement program BGMN. *J. Appl. Cryst.* **2015**, *48*, 1573–1580. [CrossRef]
16. Azhan, N.H.; Su, K.; Okimura, K.; Zaghrioui, M.; Sakai, J. Appearance of large crystalline domains in VO_2 films grown on sapphire (001) and their phase transition characteristics. *J. Appl. Phys.* **2015**, *117*, 245314. [CrossRef]
17. Ureña-Begara, F.; Crunteanu, A.; Raskin, J.-P. Raman and XPS characterization of vanadium oxide thin films with temperature. *Appl. Surf. Sci.* **2017**, *403*, 717–727. [CrossRef]
18. Shvets, P.; Maksimova, K.; Goikhman, A. Polarized Raman scattering in micrometer-sized crystals of triclinic vanadium dioxide. *J. Appl. Phys.* **2021**, *129*, 055302. [CrossRef]
19. Marezio, M.; Dernier, P.D.; McWhan, D.B.; Kachi, S. Structural aspects of the metal-insulator transition in V_5O_9. *J. Solid State Chem.* **1974**, *11*, 301–313. [CrossRef]
20. Åsbrink, S. The crystal structure of and valency distribution in the low-temperature modification of V_3O_5. The decisive importance of a few very weak reflexions in a crystal-structure determination. *Acta Cryst.* **1980**, *B36*, 1332–1339. [CrossRef]
21. Suh, T.; Kang, S.O.; Suh, I.-H. InterplanarA: A computer program for the calculation of the crystallographic interplanar angles. *Korean J. Crystallogr.* **2009**, *20*, 15–18.
22. Lee, D.; Lee, J.; Song, K.; Xue, F.; Choi, S.-Y.; Ma, Y.; Podkaminer, J.; Liu, D.; Liu, S.-C.; Chung, B.; et al. Sharpened VO_2 phase transition via controlled release of epitaxial strain. *Nano Lett.* **2017**, *17*, 5614–5619. [CrossRef] [PubMed]
23. Damodara Das, V.; Karunakaran, D. Thickness dependence of the phase transition temperature in Ag_2Se thin films. *J. Appl. Phys.* **1990**, *68*, 2105–2111. [CrossRef]

Article

Phase Composition, Hardness, and Thermal Shock Properties of AlCrTiN Hard Films with High Aluminum Content

Lijing Peng [1,2], Jun Zhang [1,2] and Xiaoyang Wang [1,2,*]

1. College of Mechanical Engineering, Shenyang University, Shenyang 110044, China
2. Key Laboratory of Research and Application of Multiple Hard Films, Shenyang 110044, China
* Correspondence: wxy927@163.com

Abstract: TiCrAlN hard films based on TiN or CrN show superior properties in terms of hardness, wear resistance, and thermal stability due to the addition of alloying elements. AlCrTiN films based on AlN may have higher thermal shock properties, but the knowledge of AlCrTiN films with high Al content has been insufficient until now. In this study, two sets of AlCrTiN hard films with different Al contents of 48 at.% and 58 at.% among metal components were prepared via multi-arc ion plating so as to investigate the effect of Al content on the phase composition, hardness, and thermal shock resistance of the films. The same microstructures, morphologies, and thicknesses of the fabricated film samples were achieved by changing the combination of cathode alloy targets and adjusting the arc source current during deposition. The surface chemical composition, cross-sectional elemental distribution, microstructure, morphology, phase composition, surface hardness, film/substrate adhesion strength, and thermal shock performance of the AlCrTiN films were examined. The obtained results reveal that the two sets of AlTiCrN hard films are face-centered cubic solid solutions with a columnar fine grain structure and a preferred growth orientation of (200) crystal plane. The hardness of the AlCrTiN films can be improved up to HV2850 by properly reducing the Al content from 58 at.% to 48 at.%. Meanwhile, the film/substrate adhesion performance is strong enough in terms of critical loads greater than 200 N. Furthermore, the AlCrTiN films maintain high thermal shock resistance at 600 °C when the Al content decreases from 58 at.% to 48 at.%. The optimal composition of the AlCrTiN hard films is 25:13:15:47 (at.%), based on the consideration of hardness, adhesion, and thermal shock cycling resistance. This optimal AlCrTiN hard film can be suggested as an option for protective coatings of hot process die tools.

Keywords: AlCrTiN hard films; multi-arc ion plating; TiN/CrN molar ratio; hardness; phase composition; thermal shock cycle

1. Introduction

Nitride hard films have been extensively studied and developed for the application to cutting tools. In ternary nitride hard films, such as AlCrN, AlTiN, and CrTiN, the atomic substitution of metal components produces substitutional solid solution $(M_xM_y)N$ nitride films, which cause lattice distortion and thus increase the film hardness. From the perspective of solid solution strengthening, when the crystal structure remains single phase without forming second phase, the larger the lattice distortion caused by the increase in the solute fraction, the more apparent the strengthening effect. Based on the stoichiometric chemical ratio of a $(M_xM_y)N$-type nitride film, the maximum solution strengthening hardness can be obtained at a molar ratio of the metal components close to 1:1 [1–3]. When accompanied by a crystal lattice distortion, the preferred crystal growth orientation and even phase composition of the film may also change [4,5].

To increase the hardness of TiN-based ternary nitride films and optimize their comprehensive properties, a third metal component is often added to form a quaternary substitutional $(Ti_xM_YM_Z)N$ nitride solid solution hard film. For example, TiAlCrN, TiAlZrN,

and TiAlNbN hard films generally exhibit high hardness and good comprehensive properties [6–11]. In consideration of the strong effect of Al on the oxidation resistance of Al-contained nitride hard films and the service requirements for nitride hard films as protective coatings of tools and molds, the thermal shock resistance of these films directly affects their actual utilization [12,13]. In fact, most studies on TiAlCrN hard films examined CrN-based and TiN-based films, while relatively few reports focused on AlCrTiN hard films with high Al contents.

Moreover, studies on AlCrN and AlTiN hard films with high Al content have shown that the AlN content up to 60–70 mol.% in $Ti_{1-x}Al_xN$ or $Cr_{1-x}Al_xN$ single solid solution phase with face-centred (fcc) cubic lattice can be retained without an occurrence of hexagonal close-packed (hcp) structure [14–16]. Additionally, the hardness of AlCrN and AlTiN hard films with fcc structure was generally larger than that with (hcp) structure. Similar results were also validated in the study of AlCrTiN hard films [8,17].

The present work aims to investigate the effects of Al content among metal components and Ti/Cr ratios on the phase composition, hardness, film/substrate adhesion strength, and thermal shock properties of AlCrTiN solid solution hard films with fcc structure. Through the deposition process design of multi-arc ion plating technology, different combinations of cathode arc source targets were selected, and the corresponding arc source current was adjusted to ensure the basic consistency of the fabricated film samples in terms of their surface morphology, cross-sectional morphology, N content distribution, and film thickness. Additionally, the deposition process was designed to ensure the Al contents among metal components in AlCrTiN hard films were less than 60 at.% so as to fabricate AlCrTiN solid solution hard films with fcc structure.

2. Materials and Methods

AlCrTiN hard films were prepared via multi-arc ion plating (MAD-4B) on high-speed steel (W18Cr4V, HRC63-64) substrates [18]. The co-deposition of AlCr and AlTi dual-arc source alloy targets (99.95% purity) with various compositions was performed to ensure the comparability of the deposited AlCrTiN hard films. The two arc source target currents were combined in different proportions to vary the film compositions, while other deposition parameters, such as the gas flow (pressure), deposition time, and deposition temperature, remained the same so as to ensure the basic consistency of the fabricated film samples in terms of their surface morphology, cross-sectional morphology, N content distribution, and film thickness. The coated samples were marked as 1#–6#, as shown in Table 1.

Table 1. Cathodic arc current combination during the deposition of AlTiCrN films (total current = 110–114 A).

Sample No.	Cathodic Arc Current (A)		Deposition Time (min)	Bias (V)
1	$Al_{70}Cr_{30}$ / 56	$Al_{70}Ti_{30}$ / 54	5 + 10 + 35	180
2	$Al_{70}Cr_{30}$ / 58	$Al_{50}Ti_{50}$ / 55	5 + 10 + 35	180
3	$Al_{50}Cr_{50}$ / 59	$Al_{64}Ti_{36}$ / 55	5 + 10 + 35	180
4	$Al_{70}Cr_{30}$ / 50	$Al_{50}Cr_{50}$ / 60	5 + 10 + 35	180
5	$Al_{50}Cr_{50}$ / 55	$Al_{64}Ti_{36}$ / 55	5 + 10 + 35	180
6	$Al_{50}Cr_{50}$ / 58	$Al_{64}Ti_{36}$ / 52	5 + 10 + 35	180

Double-target arc prebombardment was performed at a relatively high negative voltage bias (320–350 V) in an Ar atmosphere (3.0×10^{-1} Pa) for 10 min to clean the substrate and thus increase the film adhesion strength. The initial temperature of the arc bombardment process was 180 °C. After that, an alloy transition layer (deposition time: 5 min, bias voltage drop: −180 V), low N_2 flow film (deposition time: 10 min, bias voltage: −180 V), and normal N_2 flow film (deposition time: 35 min, bias voltage: −180 V) were obtained (Figure 1). The vacuum chamber temperature at the end of the entire deposition process was 200 ± 10 °C.

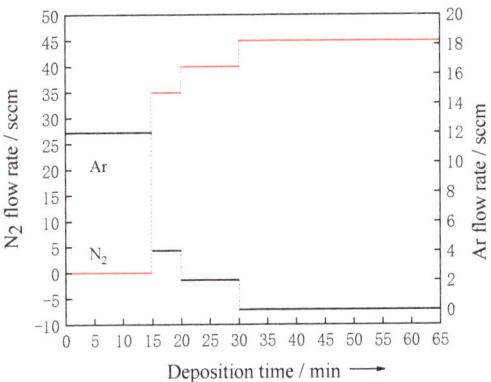

Figure 1. The flow rates of Ar and N_2 vs. deposition time. Black line: Ar, and red line: N_2.

Surface morphologies and compositions, cross-sectional morphologies, and elemental distributions of the as-deposited AlCrTiN films were observed with the aid of field-emission scanning electron microscopy (FE-SEM, S-4800) combined with energy-dispersive X-ray spectroscopy (EDS). Phase compositions of the as-deposited films were determined by X-ray diffraction (XRD, X'PertPRO) at a voltage of 40 kV, current of 40 mA, and wavelength (Cu-K α) of 0.154056 nm. The scanning angle (2θ) varied from 20° to 90° at a rate of 0.02°/min and step size of 0.05°. The obtained XRD data were analyzed by the Jade and Origin data processing software. The hardness of the as-deposited films was measured by performing Vickers micro-indentation tests (402 MVD). Three positions in the same field of view were tested at a load of 0.25 N for 20 s.

The film/substrate adhesion strengths of the prepared AlTiCrN films were determined by a WS-2005 automatic scratch tester. The maximum limit of the test load was 200 N, the loading rate was 200 N/min, the scratch speed was 4 mm/min, and the scratch length was 4 mm. A dynamic load method was used during testing, and an acoustic emission signal was recorded. The surface of each prepared sample film was examined by conducting 3–5 scratch tests.

Thermal shock tests were performed at 600 °C in a box-type electrical resistance furnace. A film sample was placed in the uniform temperature zone of the furnace for 10 min, quickly cooled to room temperature in water, and dried with hot air. The microstructure of the film layer was observed, and the test was repeated for 16 cycles.

3. Results and Discussion

3.1. Surface Morphology and Composition of AlCrTiN Hard Films

The surface morphologies of all the as-deposited films were observed by SEM. The surfaces of all coating layers were very similar and contain distributed "large particles" (liquid droplets). In general, each coating surface consisted of non-liquid and liquid droplets. These particles, with sizes below 6 μm, were formed from the micro-droplets sprayed from the cathode target surface [19,20]. Furthermore, the droplet densities on the surfaces of samples 1# and 4# were higher than those on the surfaces of other specimens, which can be attributed to the higher arc currents and Al contents of the targets used for preparing these two samples. According to the literature, Al-Cr alloy targets are more likely to produce larger particles [21,22]. In addition, some scattered micro-pits were also observed on the film surfaces, owing to the shedding of loosely attached liquid particles during the deposition process (Figure 2).

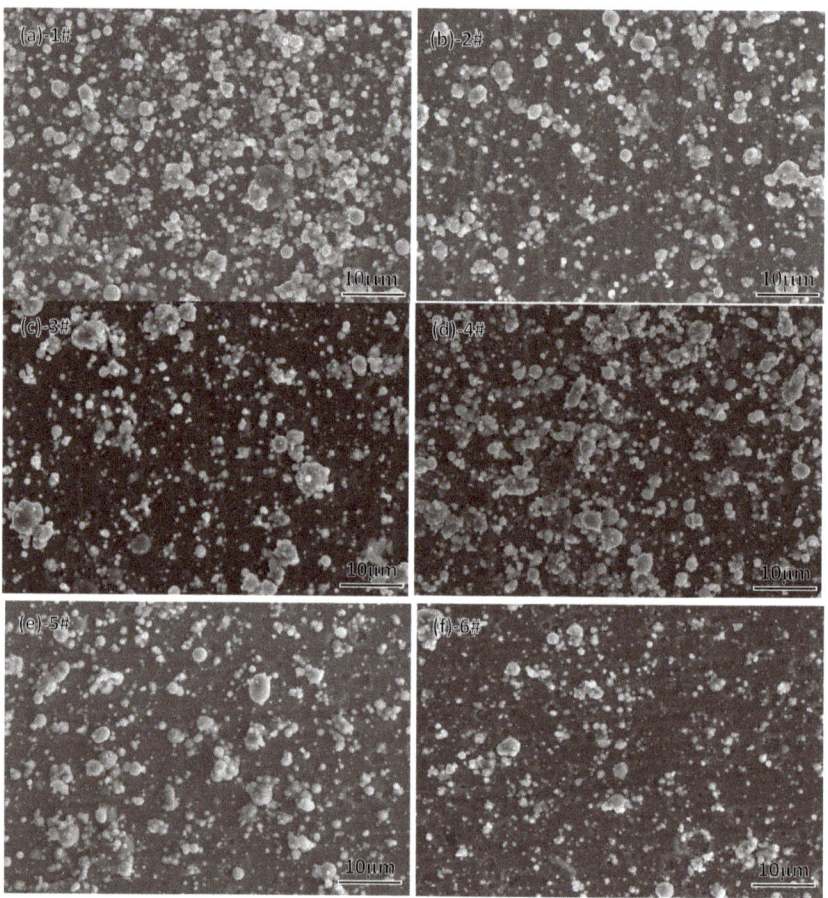

Figure 2. Surface morphologies of the deposited AlCrTiN films. The numbers (**a**–**f**) correspond to samples 1#–6#, respectively.

The surface compositions of the as-deposited films, including the droplet-free areas and large droplet particles, were determined by EDS. Because the N contents in the droplet-free areas of different film samples were nearly identical (46–47 at.%, see Figure 3a), the metal contents in the droplet-free areas could be normalized; then, an intuitive distribution triangle of the metal component contents could be established, as shown in Figure 3b. It was found that film samples 1#, 2#, 3#, and 4# exhibited almost the same Al content (58 at.%) and different Ti/Cr molar ratios, while the Al contents of samples 5# and 6# were equal to 48%. As a result, the AlCrTiN film samples with varied Al contents could be divided into two groups for the subsequent comparative study.

The droplet compositions on the film surfaces had relatively high uncertainties. Sample 3# is used as an example to illustrate this uncertainty. In the same field of view, the composition of droplet particles fluctuated significantly (Figure 4). This can be attributed to the following two reasons. First, it was closely related to the time of droplet attachment to the film surface during deposition. Thus, the droplets that attached at the early stage of the deposition process formed a surface with a composition close to that of the droplet-free areas due to the subsequent deposition of the film layer [23]. However, almost all droplets attached at the later stage of the deposition process were metallic. Second, the compositions of the arc source targets strongly influenced the droplet composition. Because different droplets could overlap on the film surface, this also increased their compositional uncertainty.

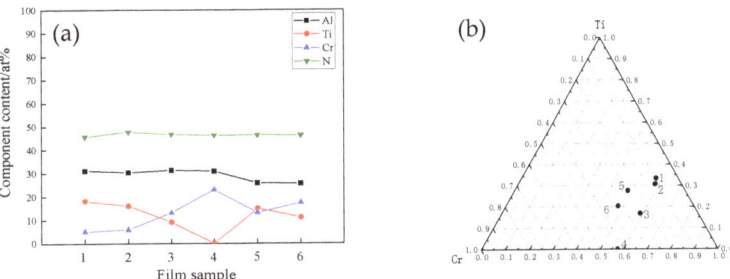

Figure 3. Chemical composition of the surface of the films. (**a**) Droplet-free film layer areas. (**b**) The corresponding normalized metal component contents.

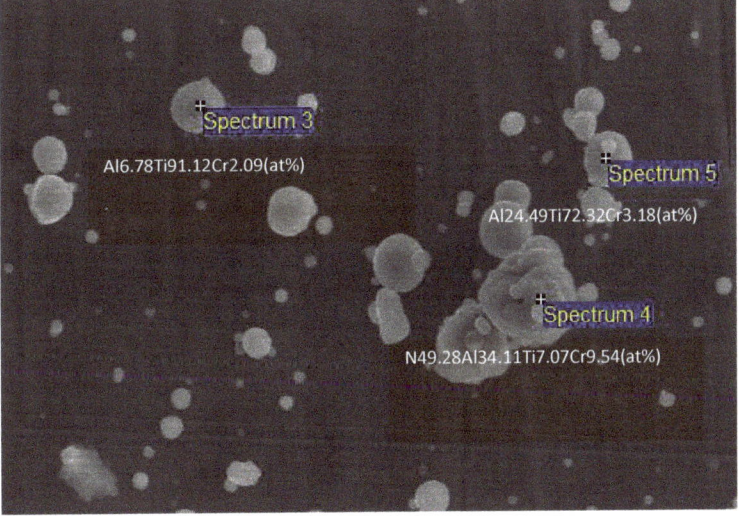

Figure 4. Chemical composition of the droplets on the surface of films.

3.2. Cross-Sectional Morphology and Elemental Distribution of AlCrTiN Hard Films

The brittle fracture cross-sectional SEM images of the as-deposited films obtained at room temperature are shown in Figure 5. The thicknesses of the films were very similar (2.1–2.3 µm) and all films exhibited dense and fine columnar crystal morphologies. Figure 6a–c present the grain growth morphology of the film surface in the droplet-free region of samples 5#, 6#, and 3#. Nearly the same fine grain growth morphology was observed in samples 1#–4#, and it was difficult to observe clear crystal grain boundaries. The possible reason is that film samples 1#–4# contained higher Al contents. In comparison, relatively clear grain boundaries were observed in samples 5# and 6#, and the grains of sample 6# were finer than those of sample 5#, with approximately half the grain cross-section size. The grain cross-section sizes of samples 1#–6# were between 30 nm and 60 nm and roughly consistent. The grain size of AlCrN films with an Al/Cr atomic ratio from 1:1 to 2.3:1, deposited by the process parameters similar to those used in the present work, was reported as about 35 nm [24]. The fine columnar crystal microstructure of multi-component nitride hard films deposited by cathodic arc ion plating at a properly high bias is a characteristic of the deposition technology [25,26].

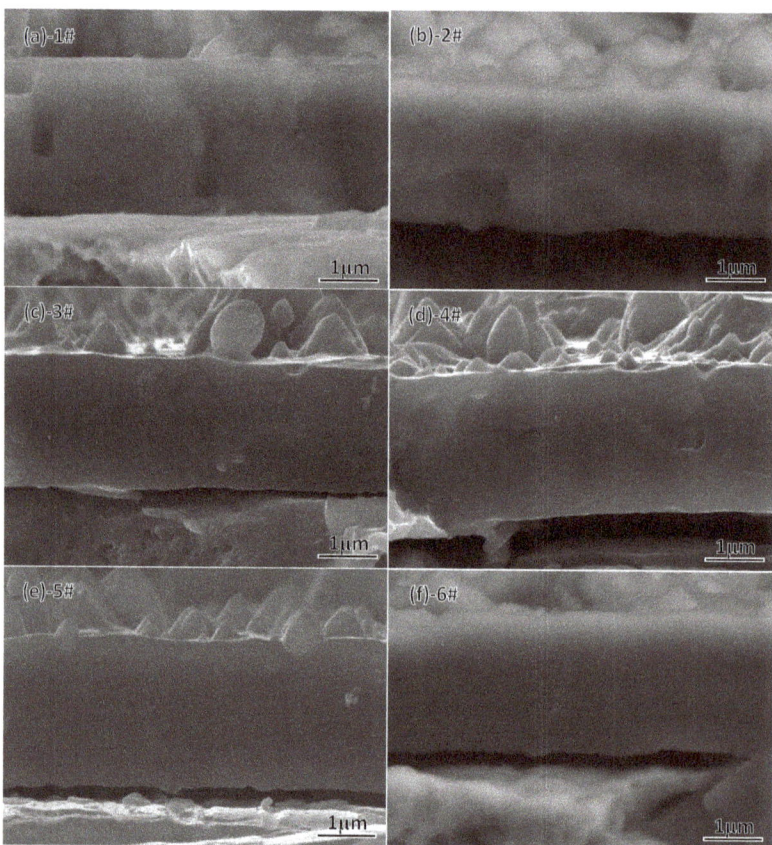

Figure 5. Cross-sectional morphologies of the films. The numbers (**a**–**f**) correspond to samples 1#–6#, respectively.

EDS line scanning was performed to determine the elemental distribution in the film growth direction. As expected, the sample coatings demonstrated similar distribution characteristics. The N content in each sample exhibited an apparent gradient in the film growth direction, while the contents of the other metal components remained almost unchanged (Figure 7).

Note that the utilized deposition technology and deposition process parameters produced different effects on the structure and properties of the multi-component nitride hard films. For example, the negative substrate bias affected the structure, composition, phase composition, and hardness of the films [27,28]. Meanwhile, the N_2 flow rate and substrate temperature influenced the N content, structure, and hardness of the film layer [29,30]. Although the compositions of the films prepared using different deposition technologies and conditions were either the same or very similar, the properties of these films differed considerably [31,32].

During deposition, the sum of the two arc currents was maintained constant at 110–114 A, and the deposition time was 5 + 10 + 35 min. Meanwhile, the negative bias voltage of the substrate as well as the argon and nitrogen flow rates at the corresponding stages of the deposition process were the same for all film samples. This ensured the nearly identical distributions of N atoms, microstructures, and thicknesses of all films; therefore, the prepared samples mainly differed in their compositions, and their phase compositions were determined by the corresponding film compositions. Finally, the film and phase compositions synergistically influenced the film performance. In summary, film composition was a single important factor affecting film characteristics, which allowed for the examination of the influences of Al content and Ti/Cr molar ratio on the phase composition, hardness, film/substrate bonding strength, and thermal shock properties of the fabricated AlCrTiN hard films.

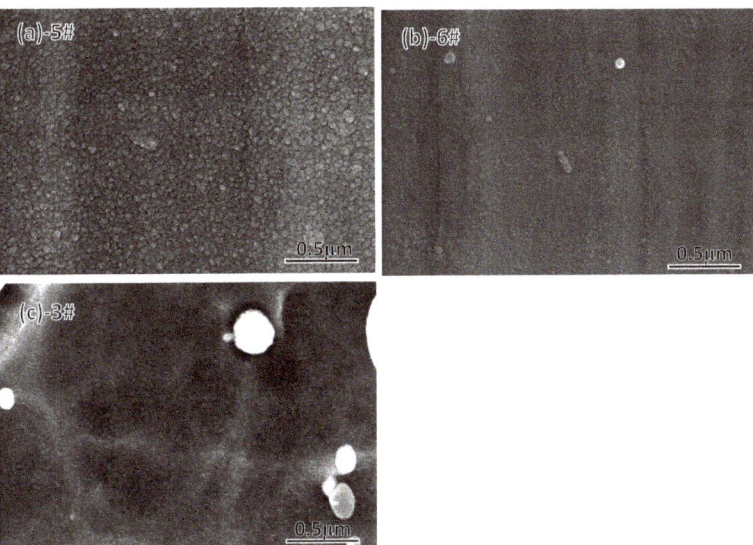

Figure 6. The grain growth morphology of the film surface in the non-droplet region; (a–c) correspond to samples 5#, 6#, and 3#, respectively.

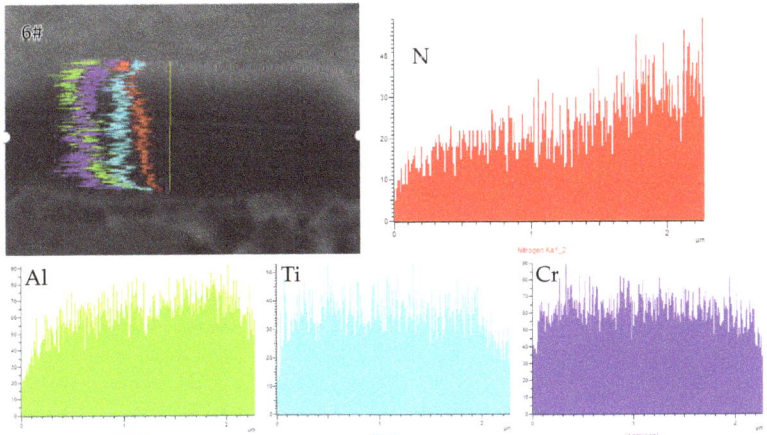

Figure 7. Cross-sectional elemental distribution in the growth direction of the films (sample 6#).

3.3. Phase Compositions of AlCrTiN Hard Films

All films were investigated by small-angle XRD, and the obtained results were analyzed by the JADE software. The XRD patterns depicted in Figure 8a,b exhibit similar characteristics in diffraction peak intensity distribution. According to the standard diffraction patterns of CrN and TiN (Figure 9a,b), the prepared AlCrTiN films are mainly composed of the face-centered cubic (fcc) AlCrTiN phase with substitutional solid solution characteristics and the preferred growth orientation of the (200) crystal plane (tagged with the solid circle symbol, in Figure 8a). Additionally, a TiAlCr alloy phase (tagged with the diamond symbol, in Figure 8a) with varied metal element concentrations, resulting from the droplets attached at the later stage of the deposition process, can be identified. Notably, the fcc AlCrTiN phase present in films 3#, 4#, and 6# exhibits a diffraction peak smoothing or separation feature (Figure 8b) due to the high Al and Cr contents [15,16]. However, unlike the phase transition tendency from fcc structure to hcp structure in $Ti_{1-x}Al_xN$ or $Cr_{1-x}Al_xN$ films as the Al content among metal components increases up to 60–70 mol.%, the fcc structure rather than hcp structure seems easily retained for AlN in the AlTiCrN films. This may be attributed to the fact that complex lattice distortion due to the addition of Ti atoms into AlCrN lattice weakens the driving force to form hcp structure of AlN phase, as shown by the XRD patterns of films 2# and 5#.

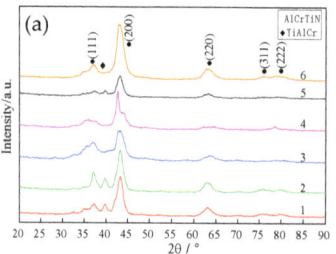 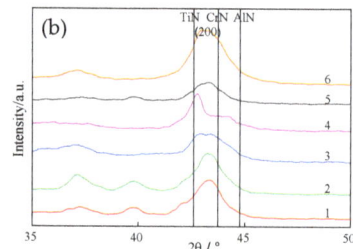

Figure 8. (**a**) XRD patterns of AlCrTiN films, (**b**) Partial enlarged detail of the XRD patterns.

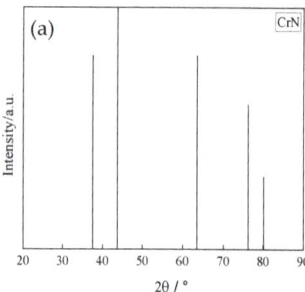 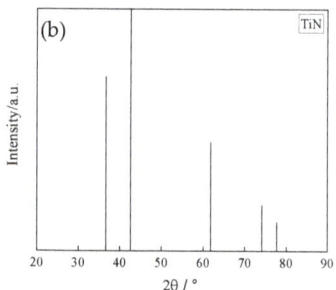

Figure 9. Standard XRD patterns of CrN and TiN. (**a**) CrN. (**b**) TiN.

The (200) crystal plane diffraction peak of film sample 4# corresponds to the Al(Cr)N fcc phase with high Al content and Cr(Al)N fcc phase with high Cr content. The Ti contents of samples 3# and 6# are similar, and these samples also exhibit high Al and Cr concentrations, indicating a separation trend. However, their Cr(Al) contents are lower than those of sample 4# because some Cr(Al) atoms are substituted by Ti atoms, which weaken the separation trend. This resulted in the formation of a single AlCrTiN phase broadening the (200) diffraction peak. Therefore, no isolated TiN, CrN, or AlN phases are present in the prepared films except for in sample 4#.

From the results obtained for samples 4#, 3#, 2#, and 1# with the same Al contents, it can be concluded that increasing the Ti/Cr ratio promotes the formation of a single fcc solid solution. The same feature was observed for samples 5# and 6#, suggesting that the AlCrTiN film easily maintains the fcc structure even at high Al contents close to 60%. Although the Al contents in all films (except for sample 4#) are as high as 48%–58%, their diffraction peaks remain between the standard diffraction peaks of the TiN and CrN phases, indicating that the peaks of the (200) crystal plane obtained for these films are shifted to smaller angles.

As shown in Figure 3b, the AlCrTiN film can be seen to be a substitutional solid solution formed by the TiN, CrN, and AlN phases, and their respective proportions are replaced by the normalized proportions of metal element concentrations. The ideal lattice constants of the fcc solid solution of each group of the fabricated film samples were calculated according to Vegard's law. Additionally, considering the characteristic diffraction peak position of the (200) crystal plane of the AlCrTiN film, the lattice constant of the fcc (AlCrTi)N solid solution phase was directly determined using the JADE software. The obtained results are shown in Figure 10. The experimentally measured lattice constants of all films (except for sample 4#) are significantly higher than the values calculated by Vegard's law because TiN, CrN, and AlN form a complex fcc solid solution, resulting in large lattice distortions and the formation of vacancy defects. Because the conducted film deposition was a non-equilibrium process, point defects and residual thermal stress were generated in the films, increasing the lattice constant [33].

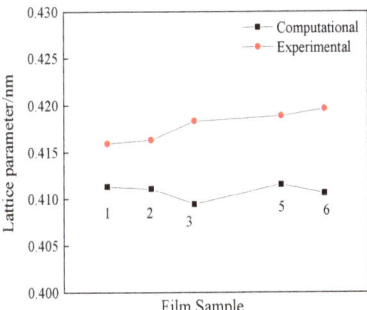

Figure 10. Lattice constant of the films obtained from the experiments and calculations using Vegard's law. Notes: 1. excluding film 4#; 2. the 2θ angle of (200) crystal plane of films 3# and 6# is an estimated value.

The texture characteristics of a multi-component nitride film are strongly related to the utilized deposition technology and process parameters. The $Cr_{1-x}Al_xN$ film prepared by multi-arc ion plating technology, which is similar to the fabrication method used in the present study, exhibited similar properties (a B1-NaCl type fcc structure was formed below $x = 0.6$, and its preferred growth orientations corresponded to the (200) and (111) crystal planes) [34]. However, the $Al_xCr_{1-x}N$ film produced by unbalanced magnetron sputtering with two sputtering sources at $0.31 \leq x \leq 0.71$ had a preferred orientation of the single (111) crystal plane [35].

In this work, the single composition change did not change the phase composition or crystal structure of the AlCrTiN films, owing to the consistency of the deposition process. All films (except for 4#) were composed of the single AlCrTiN substitution fcc solid solution phase and had the same preferred growth orientation corresponding to the (200) crystal plane. Therefore, it can be concluded that the performance difference of these films originated from composition changes. This effect mainly involved solution strengthening caused by lattice distortion.

3.4. Hardness and Adhesion of AlCrTiN Hard Films

The microhardness test data obtained for all film layers of droplet-free areas are listed in Table 2. It shows that the hardness values of AlCrTiN samples 1#–4# with equal molar ratios of AlN (~58%) are lower than those of the AlCrN and AlTiN films with the same AlN molar ratio [16,35]. These results indicate that the hardness of the AlCrTiN films formed by adding Ti or Cr atoms to AlCrN or AlTiN films with high Al contents is relatively low.

However, with a decrease in the molar ratio of AlN to 48%, the hardness values of the AlCrTiN films 5# and 6# increase as compared with those of films 2# and 3#.

Table 2. The microhardness of AlCrTiN films.

Sample No.	AlN/mol %	TiN/mol %	CrN/mol %	TiN/CrN	Microhardness (HV)	Adhesion (N)
1	57.14	33.39	9.44	3.54	2680 ± 200	>150
2	58.14	30.67	11.2	2.74	2650 ± 200	>200
3	58.67	16.88	24.47	0.69	2620 ± 200	>200
4	57.40	0	42.6	0	2750 ± 200	>130
5	48.06	27.6	24.36	1.13	2850 ± 200	>200
6	47.54	20.33	32.13	0.63	2730 ± 200	>200

Under the premise of maintaining the fcc structure of a ternary substitutional solid solution nitride film (such as AlCrN, TiAlN, and TiCrN), as the content of the second metal component increases, the lattice distortion will increase, and then the hardness of the film layer tends to increase [3,34,36]. The hardness maximum value is reached when the metal component ratio is close to 1:1.

The present AlTiCrN substitutional solid solution films were formed by adding Ti atoms to AlCrN fcc structure and replacing some of the Al and Cr atoms because the ratio of N/Metal is the same and almost equal to 1:1. In this case, three solid solution combinations were produced: a combination of AlN and TiN, a combination of AlN and CrN (both of which have high AlN fractions), and a combination of TiN and CrN. Therefore, the hardness of the AlTiCrN solid solution film was reduced because the AlCrN and AlTiN films with high Al contents exhibited low hardness values. Only when the addition of Ti atoms reached a certain fraction among the metal components and the AlTiCrN solid solution was gradually transformed into a combination of AlN/TiN and AlN/CrN, both with a small concentration difference, did the film hardness increase, as shown by films 5# and 6#.

According to previous studies, the hardness of the $Cr_{0.66}Al_{0.34}N$ ternary film is 28.5 GPa. With the addition of the Ti component at f_{Ti} = 5% and 10%, the hardness values of the resulting $Cr_{65}Ti_5Al_{30}N$ and $Cr_{61}Ti_{10}Al_{29}N$ films increase to 34 and 40 GPa, respectively. When f_{Ti} is further increased, the hardness begins to decrease slowly. When the Ti content reaches 64%, the hardness of the $Cr_{26}Ti_{64}Al_{10}N$ film decreases to 29 GPa [37]. However, the hardness of the $Ti_{0.17}Al_{0.53}Cr_{0.30}N$, $Ti_{0.23}Al_{0.36}Cr_{0.4}N$, and $Ti_{0.46}Al_{0.20}Cr_{0.34}N$ films prepared using different combinations of the TiAl, Cr, and Al targets monotonically increases with a decrease in the Al content and increase in the Ti content [17]. The work to investigate the effect of bias on the composition and hardness of TiAlCrN films shows that as the bias increases from 100 V, 150 V, 200 V to 250 V, the composition of the films is successively changed into $Ti_{23}Al_{18}Cr_{20}N$, $Ti_{27}Al_{16}Cr_{21}N$, $Ti_{20}Al_{14}Cr_{29}N$, $Ti_{23}Al_{12}Cr_{31}N$, and the hardness is successively changed into HV2800, HV5100, HV4650, HV4650V [38]. Thus, it can be concluded that the hardness values of the TiN-based TiAlCrN and CrN-based CrTiAlN films are generally higher than that of the AlN-based AlTiCrN film.

In addition, the N/metal ratio of nitride hard films directly affects their hardness, which decreases when this ratio deviates significantly from the stoichiometric ratio of 1:1 [39,40]. Therefore, the N content in the hardening film layer is generally maintained between 45 and 52 at.%, which is close to the ideal chemical ratio required to maximize the hardness of the nitride film. The N contents in the hardening surface layers of the as-deposited AlTiCrN solid solution films prepared in this study amounted to 47%–48%, which ensured their maximum hardness at various ratios of the metal components.

The film/substate adhesive strength values of the AlCrTiN hard films (Table 2) suggest that the adhesion between the film and substrate is relatively strong, except for film samples 1# and 4# (the other films layers withstood critical loads exceeding 200 N). Previous studies have shown that designing the N gradient distribution in nitride hard films is an effective method for improving the film/substrate adhesion performance [41,42]. For example, the adhesive strength of a (TiAlNb)N film with N gradient distribution prepared was much larger than that of a (TiAlNb)N film without an alloy transition layer and N gradient distribution [11,43].

In the present study, the ion bombardment of the high-speed steel substrate prior to film deposition and subsequent deposition of the alloy transition layer prevented the separation of the film from the substrate and increased the film adhesive strength. Optimizing the gradient distribution of the N element in the film growth direction inhibits the accumulation of growth stress and produces a gradient of the thermal expansion coefficient, which helps improve both the adhesion [41,42] and thermal shock performance of the film.

3.5. Thermal Shock Cycling Performance of AlCrTiN Hard Films

The cutting tools and dies coated with nitride hard films are generally required to have long service lives and good machining properties under hot and cold cycling conditions. Therefore, the corresponding hard films must possess high thermal shock resistances.

The fabricated AlCrTiN hard films were subjected to thermal shock cycles at 600 °C, and the changes in the surface morphology were observed after each cycle. After eight cycles, no significant changes in the film surfaces were detected by SEM with a 2000× magnification. The surface changes of two samples, 1# with more droplets and 5# with fewer droplets, are given as examples, as shown in Figure 11. To observe the film surface more carefully, the magnification was increased to 10,000×. During the entire eight thermal shock cycles, the observable change in the surface morphology was that the surface of droplet particles became coarse, and some very small particles appeared. In contrast, negligible morphological changes occurred in the droplet-free areas. This indicates that the surface oxidation rate of the droplets was significantly higher than that of the droplet-free area. After the eighth cycle, neither cracking or shedding of droplets nor cracks on the surface of the droplet-free area were detected in any of the samples. Figure 12a,b show the surface changes of samples 2# and 3#, respectively.

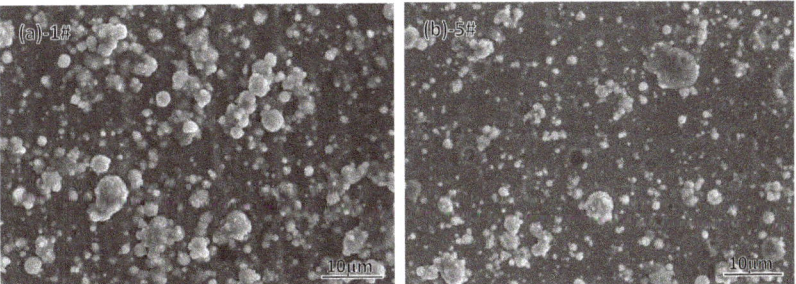

Figure 11. Surface morphologies of AlTiCrN hard films after the 8th thermal shock cycle. (**a**) Sample 1#. (**b**) Sample 5#.

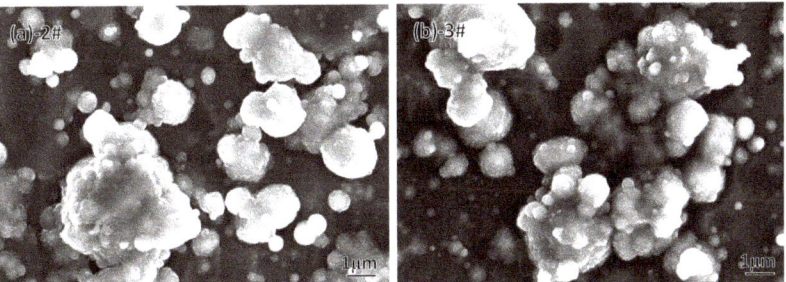

Figure 12. Surface morphologies of film samples 2# and 3# after 8th thermal shock cycle. (**a**) Sample 2#. (**b**) Sample 3#.

After the 16th cycle, inconsistent changes were observed on the film surfaces with different compositions (Figure 13). For films 1# and 4#, no obvious changes in the film surface morphologies (the droplet-free areas) were observed, as shown by the red circle in Figure 13a,d. However, a few of cracks formed on the surfaces of droplet particles. For films 2# and 3#, the obvious growth of the cross-section grain size of the film layer (the droplet-free areas) was observed up to about 200 nm and 300 nm, respectively, as shown by the red circle in Figure 13b,c. Meanwhile, a few cracks were observed on droplet particles. For films 5# and 6#, on the film surface of the droplet-free areas, the grain boundaries became blurred and the cross-section size of grain in film 5# seemed to be larger than that in film 6#, as shown by the red circle in Figure 13e,f. Additionally, no visible cracks formed on the surfaces of droplet particles. In general, all films exhibited no cracks in the droplet-free area through 16 thermal shock cycles. This means that the prepared film samples possessed high thermal shock resistances.

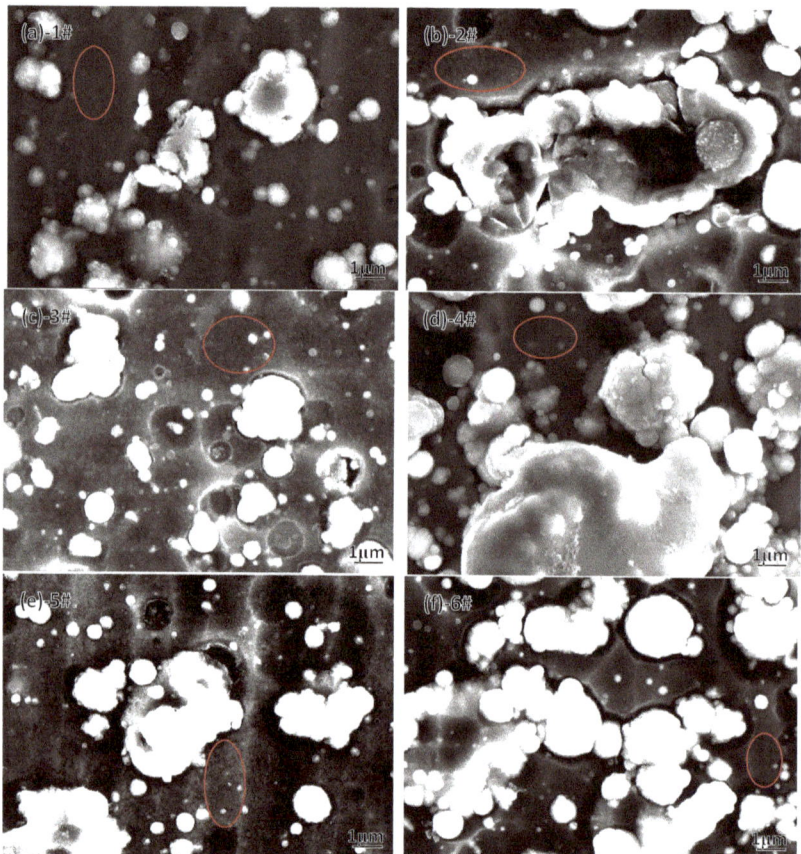

Figure 13. Surface morphologies of film samples after 16 thermal shock cycles. The numbers (a–f) correspond to sample numbers 1#–6#, respectively. Red circle region: droplet-free area.

The Al element has a relatively high oxidation resistance because it easily forms a dense oxide (Al_2O_3) layer that covers the film surface during thermal shock, thus preventing oxygen atoms from further penetrating the film and resulting in the oxidation of the film layer [44]. However, the Cr element can promote the oxidation of Al atoms and accelerate the formation of Al_2O_3 [45]. Therefore, a film with high Al content typically exhibits high thermal shock resistance. In the present study, all film layers of the droplet-free areas with high Al and Cr contents demonstrated high thermal shock resistance. Because the thermal shock cycling temperature was 600 °C, which is lower than the preferred oxidation temperature of Al [46–48], films 1#–4# with higher Al contents, in comparison with films 5# and 6#, did not exhibit obvious advantages in terms of thermal shock resistance.

4. Conclusions

In this study, two sets of AlCrTiN hard films with different Al contents among metal components were prepared. All of the films exhibited almost identical surface morphologies, cross-sectional microstructures and morphologies, thicknesses, and N content distributions because the combination of alloy targets and parameters of the deposition process were adjusted for each sample. The main conclusions from the obtained results are summarized below.

1. All AlTiCrN hard films (i.e., droplet-free areas, the same below) consist of the fcc solid solutions with a columnar fine microstructure and the preferred growth orientation of (200) crystal plane at high Al contents from 48 up to 58 at.%. AlCrTiN films easily maintains the fcc structure even at high Al contents up to 58 at.% due to the complex lattice distortion.
2. The hardness of AlTiCrN films with an Al content of 58 at.% is significantly lower than those of TiCrAlN and CrTiAlN hard films with high Ti and Cr contents, and varying the Ti/Cr ratio does not increase hardness of the AlTiCrN films.
3. The hardness of AlCrTiN films can be improved up to HV2850 by properly reducing the Al content from 58 at.% to 48 at.%. Meanwhile, the film/substrate adhesion performance is strong enough in terms of critical loads greater than 200 N.
4. AlCrTiN films maintain high thermal shock resistance at 600 °C even when the Al content decreases from 58 at.% to 48 at.%. In the droplet-free area of the film surface, no crack appeared through 16 thermal shock cycles. The thermal shock failure of the films is typically manifested as a rupture and droplet particles falling off.
5. The optimal composition of AlCrTiN hard films is 25:13:15:47 (at.%), which is determined from the results of hardness, adhesion, and thermal shock cycling resistance measurements. This optimal AlCrTiN hard film can be used as an option for protective coatings of hot-pressure die tools.

Author Contributions: Conceptualization, methodology, J.Z.; investigation, L.P.; resources, X.W.; data curation, writing—original draft preparation, L.P.; writing—review and editing, X.W.; funding acquisition, X.W. All authors have read and agreed to the published version of the manuscript.

Funding: This research was funded by Liaoning Province Doctor Start-up Fund, grant number 2021-BS-276.

Institutional Review Board Statement: Not applicable.

Informed Consent Statement: Not applicable.

Data Availability Statement: Not applicable.

Conflicts of Interest: The authors declare no conflict of interest.

References

1. Hörling, A.; Hultman, L.; Odén, M.; Sjölén, J.; Karlsson, L. Mechanical properties and machining performance of $Ti_{1-x}Al_xN$-coated cutting tools. *Surf. Coat. Technol.* **2004**, *191*, 384–392. [CrossRef]
2. Lamni, R.; Sanjinés, R.; Parlinska-Wojtan, M.; Karimi, A.; Lévy, F. Microstructure and nanohardness properties of Zr-Al-N and Zr-Cr-N thin films. *J. Vac. Sci. Technol. A* **2005**, *23*, 593–598. [CrossRef]
3. Niu, E.W.; Li, L.; Lv, G.H.; Chen, H.; Li, X.Z.; Yang, X.Z.; Yang, S.Z. Characterization of Ti-Zr-N films deposited by cathodic vacuum arc with different substrate bias. *Appl. Surf. Sci.* **2008**, *254*, 3909–3914. [CrossRef]
4. Falub, C.V.; Karimi, A.; Ante, M.; Kalss, W. Interdependence between stress and texture in arc evaporated Ti-Al-N thin films. *Surf. Coat. Technol.* **2007**, *201*, 5891–5898. [CrossRef]
5. Kimura, A.; Hasegawa, H.; Yamada, K.; Suzuki, T. Metastable $Ti_{1-x}Al_xN$ films with different Al content. *J. Mater. Sci. Lett.* **2000**, *19*, 601–602. [CrossRef]

6. Donohue, L.A.; Cawley, J.; Brooks, J.S.; Münz, W.D. Deposition and characterization of TiAlZrN films produced by a combined steered arc and unbalanced magnetron sputtering technique. *Surf. Coat. Technol.* **1995**, *74*, 123–134. [CrossRef]
7. Zhang, J.; Guo, W.; Zhang, Y.; Guo, Q.; Wang, C.; Zhang, L. Mechanical properties and phase structure of (TiAlZr)N films deposited by multi arc ion plating. *Thin Solid Film.* **2009**, *517*, 4830–4834. [CrossRef]
8. Yamamoto, K.; Sato, T.; Takahara, K.; Hanaguri, K. Properties of (Ti,Cr,Al)N coatings with high Al content deposited by new plasma enhanced arc-cathode. *Surf. Coat. Technol.* **2003**, *174*, 620–626. [CrossRef]
9. Santana, A.E.; Karimi, A.; Derflinger, V.H.; Schütze, A. Microstructure and mechanical behavior of TiAlCrN multilayer thin films. *Surf. Coat. Technol.* **2003**, *177*, 334–340. [CrossRef]
10. Hsu, C.H.; Chen, K.L.; Lin, Z.H.; Su, C.Y.; Lin, C.K. Bias effects on the tribological behavior of cathodic arc evaporated CrTiAlN coatings on AISI 304 stainless steel. *Thin Solid Film.* **2010**, *518*, 3825–3829. [CrossRef]
11. Zhang, J.; Yin, L.Y. Microstructure and mechanical properties of (Ti,Al,Nb)N hard films with N-gradient distributions. *Thin Solid Film.* **2015**, *584*, 141–145. [CrossRef]
12. PalDey, S.; Deevi, S.C. Single layer and multilayer wear resistant coatings of (Ti,Al)N: A review. *Mater. Sci. Eng. A* **2003**, *342*, 58–79. [CrossRef]
13. Greczynski, G.; Hultman, L.; Odén, M. X-ray photoelectron spectroscopy studies of $Ti_{1−x}Al_xN$ ($0 \leq x \leq 0.83$) high temperature oxidation: The crucial role of Al concentration. *Surf. Coat. Technol.* **2019**, *374*, 923–934. [CrossRef]
14. Pemmasani, S.P.; Valleti, K.; Gundakaram, R.C.; Rajulapati, K.V.; Mantripragada, R.; Koppoju, S.; Joshi, S.V. Effect of microstructure and phase constitution on mechanical properties of $Ti_{1−x}Al_xN$ coatings. *Appl. Surf. Sci.* **2014**, *313*, 936–946. [CrossRef]
15. Wang, D.; Lin, S.S.; Liu, L.Y.; Xue, Y.N.; Yang, H.Z.; Jiang, B.L.; Zhou, K.S. Effect of Bias Voltage on Microstructure and Erosion Resistance of CrAlN Coatings Deposited by Arc Ion Plating. *Rare Met. Mater. Eng.* **2020**, *49*, 2583–2590.
16. Hasegawa, H.; Suzuki, T. Effects of second metal contents on microstructure and micro-hardness of ternary nitride films synthesized by cathodic arc method. *Surf. Coat. Technol.* **2004**, *188–189*, 234–240. [CrossRef]
17. Vattanaprateep, N.; Panich, N.; Surinphong, S.; Tungasmita, S.; Wangyao, P. Structural and Mechanical Properties of Nanostructured TiAlCrN Thin Films Deposited by Cathodic Arc Deposition. *High Temp. Mater. Proc.* **2013**, *32*, 107–111. [CrossRef]
18. Zhang, J.; Peng, L.; Wang, X.; Liu, D.; Wang, N. Effects of Zr/(Zr+Ti) Molar Ratio on the Phase Structure and Hardness of $Ti_xZr_{1−x}N$ Films. *Coatings* **2021**, *11*, 1342. [CrossRef]
19. Harris, S.G.; Doyle, E.D.; Wong, Y.C.; Munroe, P.R.; Cairney, J.M.; Long, J.M. Reducing the macroparticle content of cathodic arc evaporated TiN coatings. *Surf. Coat. Technol.* **2003**, *183*, 283–294. [CrossRef]
20. Shiao, M.H.; Shieu, F.S. Formation of macroparticles in arc ion-plated nitride coatings. *J. Vac. Sci. Technol. A Vac. Surf. Film.* **2001**, *19*, 703–705. [CrossRef]
21. Wan, X.S.; Zhao, S.S.; Yang, Y.; Gong, J.; Sun, C. Effects of nitrogen pressure and pulse bias voltage on the properties of Cr-N coatings deposited by arc ion plating. *Surf. Coat. Technol.* **2009**, *204*, 1800–1810. [CrossRef]
22. Vyskocil, J.; Musil, J. Cathodic arc evaporation in thin film technology. *Vac. Sci. Technol.* **1992**, *A10*, 1740–1748. [CrossRef]
23. Shiao, M.H.; Shieu, F.S. A formation mechanism for the macroparticles in arc ion-plated TiN films. *Thin Solid Films* **2001**, *386*, 27–31. [CrossRef]
24. Tritremmel, C.; Daniel, R.; Lechthaler, M.; Polcik, P.; Mitterer, C. Influence of Al and Si content on structure and mechanical properties of arc evaporated Al-Cr-Si-N thin films. *Thin Solid Film.* **2013**, *534*, 403–409. [CrossRef]
25. Kuczyk, M.; Krülle, T.; Zawischa, M.; Kaspar, J.; Zimmer, O.; Leonhardt, M.; Leyens, C.; Zimmermann, M. Microstructure and mechanical properties of high entropy alloy nitride coatings deposited via direct current cathodic vacuum arc deposition. *Surf. Coat. Technol.* **2022**, *448*, 128916–128930. [CrossRef]
26. Chen, W.L.; Zheng, J.; Lin, Y.; Kwon, S.; Zhang, S.L. Comparison of AlCrN and AlCrTiSiN coatings deposited on the surface of plasma nitrocarburized high carbon steels. *Appl. Surf. Sci.* **2015**, *332*, 525–532. [CrossRef]
27. Cai, F.; Chen, M.; Li, M.; Zhang, S. Influence of negative bias voltage on microstructure and property of Al-Ti-N films deposited by multi-arc ion plating. *Ceram. Int.* **2017**, *43*, 3774–3783. [CrossRef]
28. Liu, W.; Li, A.; Wu, H.; He, R.; Huang, J.; Long, Y.; Deng, X.; Wang, Q.; Wang, C.; Wu, S. Effects of bias voltage on microstructure, mechanical properties, and wear mechanism of novel quaternary (Ti, Al, Zr)N coating on the surface of silicon nitride ceramic cutting tool. *Ceram. Int.* **2016**, *42*, 17693–17697. [CrossRef]
29. Tang, J.F.; Lin, C.Y.; Yang, F.C.; Chang, C.L. Influence of Nitrogen Content and Bias Voltage on Residual Stress and the Tribological and Mechanical Properties of CrAlN Films. *Coatings* **2020**, *10*, 546. [CrossRef]
30. Larijani, M.M.; Tabrizi, N.; Norouzian, S.; Jafari, A.; Lahouti, S.; Hosseini, H.H.; Afshari, N. Structural and mechanical properties of ZrN films prepared by ion beam sputtering with varying N_2/Ar ratio and substrate temperature. *Vacuum* **2006**, *81*, 550–555. [CrossRef]
31. Donohue, L.A.; Cawley, J.; Brooks, J.S. Deposition and characterisation of arc-bond sputter Ti_xZr_yN coatings from pure metallic and segmented targets. *Surf. Coat. Technol.* **1995**, *72*, 128–138. [CrossRef]
32. Le Bourhis, E.; Goudeau, P.; Staia, M.H.; Carrasquero, E.; Puchi-Cabrera, E.S. Mechanical properties of hard AlCrN-based coated substrates. *Surf. Coat. Technol.* **2009**, *203*, 2961–2968. [CrossRef]
33. Elstner, F.; Gautier, C.; Moussaoui, H.; Piot, O.; Machet, J. A comparative study of structure and residual stress in chromium nitride films deposited by vacuum arc evaporation, ion plating, and DC magnetron sputtering. *Phys. Stat. Sol.* **2010**, *158*, 505–521. [CrossRef]

34. Hasegawa, H.; Kawate, M.; Suzuki, T. Effects of Al contents on microstructures of $Cr_{1-x}Al_xN$ and $Zr_{1-x}Al_xN$ films synthesized by cathodic arc method. *Surf. Coat. Technol.* **2005**, *200*, 2409–2413. [CrossRef]
35. Kim, G.S.; Lee, S.Y. Microstructure and mechanical properties of AlCrN films deposited by CFUBMS. *Surf. Coat. Technol.* **2006**, *201*, 4361–4366. [CrossRef]
36. Boxman, R.L.; Zhitomirsky, V.N.; Grimberg, I.; Rapoport, L.; Goldsmith, S.; Weiss, B.Z. Structure and hardness of vacuum arc deposited multi-component nitride coatings of Ti, Zr and Nb. *Surf. Coat. Technol.* **2000**, *125*, 257–262. [CrossRef]
37. Lin, J.; Zhang, X.; Ou, Y.; Wei, R. The structure, oxidation resistance, mechanical and tribological properties of CrTiAlN coatings. *Surf. Coat. Technol.* **2015**, *277*, 58–66. [CrossRef]
38. Zhang, J.; Lv, H.; Cui, G.; Jing, Z.; Wang, C. Effects of bias voltage on the microstructure and mechanical properties of (Ti,Al,Cr)N hard films with N-gradient distributions. *Thin Solid Film.* **2011**, *519*, 4818–4823. [CrossRef]
39. Tay, B.K.; Shi, X.; Yang, H.S.; Tan, H.S.; Chua, D.; Teo, S.Y. The effect of deposition conditions on the properties of TiN thin films prepared by filtered cathodic vacuum-arc technique. *Surf. Coat. Technol.* **1999**, *111*, 229–233. [CrossRef]
40. Sundgren, J.-E. Structure and propertiesof TiN coatings. *Thin Solid Film.* **1985**, *128*, 21–44. [CrossRef]
41. Chen, L.; Wang, S.Q.; Du, Y.; Li, J. Microstructure and mechanical properties of gradient Ti(C,N) and TiN/Ti(C,N) multilayer PVD coatings. *Mater. Sci. Eng. A* **2007**, *478*, 336–339. [CrossRef]
42. Dobrzański, L.A.; Żukowska, L.W.; Mikuła, J.; Gołombek, K.; Pakuła, D.; Pancielejko, M. Structure and mechanical properties of gradient PVD coatings. *J. Mater. Process.Technol.* **2008**, *201*, 310–314. [CrossRef]
43. Yin, L.Y.; Zhang, J. Microstructure and mechanical properties of (TiAlNb)N films. *Adv. Mater. Res.* **2014**, *812*, 915–916. [CrossRef]
44. Polcar, T.; Cavaleiro, A. High-temperature tribological properties of CrAlN, CrAlSiN and AlCrSiN coatings. *Surf. Coat. Technol.* **2011**, *206*, 1244–1251. [CrossRef]
45. Kawate, M.; Hashimoto, A.K.; Suzuki, T. Oxidation Resistance of CrAlN and TiAlN Films. *Surf. Coat. Technol.* **2003**, *165*, 163–167. [CrossRef]
46. Endrino, J.L.; Fox-Rabinovich, G.S.; Gey, C. Hard AlTiN, AlCrN PVD Coatings for Machining of Austenitic Stainless Steel. *Surf. Coat. Technol* **2006**, *200*, 6840–6845. [CrossRef]
47. Fox-Rabinovich, G.S.; Beake, B.D.; Endrino, J.L.; Veldhuis, S.C.; Parkinson, R.; Shuster, L.S.; Migranov, M.S. Effect of Mechanical Properties Measured at Room and Elevated Temperatures on the Wear Resistance of Cutting Tools with TiAlN and AlCrN Coatings. *Surf. Coat. Technol.* **2006**, *200*, 5738–5742. [CrossRef]
48. Jäger, N.; Meindlhumer, M.; Zitek, M.; Spor, S.; Hruby, H.; Nahif, F.; Julin, J.; Rosenthal, M.; Keckes, J.; Mitterer, C.; et al. Impact of Si on the high-temperature oxidation of AlCr(Si)N coatings. *J. Mater. Sci. Technol.* **2022**, *100*, 91–100. [CrossRef]

Disclaimer/Publisher's Note: The statements, opinions and data contained in all publications are solely those of the individual author(s) and contributor(s) and not of MDPI and/or the editor(s). MDPI and/or the editor(s) disclaim responsibility for any injury to people or property resulting from any ideas, methods, instructions or products referred to in the content.

MDPI
St. Alban-Anlage 66
4052 Basel
Switzerland
www.mdpi.com

Coatings Editorial Office
E-mail: coatings@mdpi.com
www.mdpi.com/journal/coatings

Disclaimer/Publisher's Note: The statements, opinions and data contained in all publications are solely those of the individual author(s) and contributor(s) and not of MDPI and/or the editor(s). MDPI and/or the editor(s) disclaim responsibility for any injury to people or property resulting from any ideas, methods, instructions or products referred to in the content.

www.ingramcontent.com/pod-product-compliance
Lightning Source LLC
LaVergne TN
LVHW070603100526
838202LV00012B/551